THE OFF GRID SOLAR POWER BIBLE

[8 in 1] The Complete How-To Manual for Designing, Installing, and Maintaining Solar Energy Systems for RVs, Cabins, Boats, and Tiny Homes, 2026 Edition

---

Edward Allen

Table of Contents

Introduction

The issue that the world is currently experiencing has gotten worse during the past century. The atmosphere of the Earth has been under siege since the industrial revolution, and greenhouse gases have been eliminating airborne safe gases. The loss of ozone is only one of numerous setbacks that have made life on Earth more challenging than ever.

Because of the development of new technologies, the need for fossil fuels is more important than ever. Making matters worse, fossil fuel supplies are depleting while simultaneously contributing to Earth's destruction. Along with the rest of the globe, third world nations are currently becoming more dependent on fossil fuels. Each meaningful attempt is founded on a dependence on fossil fuels due to the development of water purifiers and equipment for the mass distribution of goods.

Green energy is an ideology that unites all nations in efforts to start building a better planet. Green energy not only offers the potential for renewable energy but will also assist in the restoration of the health of our planet. The creation of energy without the use of pollutants is known as "green energy," in case you've genuinely been living under a rock. A few examples of renewable energy sources are hydroelectricity, geothermal, and wind.

Each, however, has challenges. Windmills, which are only effective in windy areas, are used to generate wind energy. In the event that there is no wind, no energy is generated. Geothermal energy uses the heat of the Earth to generate power, but it is frequently expensive and needs a large hole in the Earth to extract its energy from. Large rivers are frequently used to generate hydroelectricity, but these are rare in desert regions. Not all of these are widely accessible everywhere. A different approach is available, though, and its use is rising quickly.

One of the most widely used strategies for generating energy with the least amount of carbon footprint is solar. Solar technologies harness one of our most renewable resources to date—the sun—instead of spewing hundreds of pounds of greenhouse gases into the atmosphere. Although solar energy is nothing new, there have been tremendous advances in its utilization over time.

What can you therefore do about it? You can now convert your own home, RV, van, car, or boat into a solar-generating machine for a lot less money than you could 10 years ago thanks to the development of more affordable materials. Additionally, there are more modifications on the market, so experts will charge you a fair amount to install solar panels for you. But if you're resourceful and daring, you can assemble everything yourself.

It's normal to think that you lack the abilities needed to finish the task, but nothing could be further from the reality. This book was actually written to help you design your own installations. This book not only explains how to build your own solar panel arrangement, but it also gives you average costs and wiring diagrams to help you through the process.

You may learn how to construct mobile solar power stations for boats, RVs, automobiles, and dollies in this book. Basically, you should be able to recharge batteries and power electronics both on land and in the water. It is up to you to explore the theoretical limits of adding additional electricity to any project, local or remote.

Why People Live Off-Grid?

Off-grid living is an option for some people for a variety of reasons. Others desire the cost-cutting advantages of being cut off from the public grid. If there are power outages in your neighborhood, you can live off the grid to have energy security. Off-grid power refers to electricity that is not connected to the municipal grid and is typically produced using renewable energy sources.

The three main off-grid power generation technologies are wind, solar, and micro-hydro. In contrast to the alternating current provided by electrical grids, these energy sources convert their power into direct current electricity. Here are the three primary sorts of energy sources you need to be familiar with if you're considering living off the grid. These renewable energy sources all have the same and most desirable quality: they are environmentally benign. They leave no carbon footprint behind. Even though the initial setup expenses for renewable energy sources can be very costly, they are cost-effective in the long term.

Solar Energy

The majority of individuals who live off the grid rely on solar electricity, which is by far the most prevalent renewable energy source. Solar power is more affordable than other options, such as generators, if you live off the grid. You should use as little energy and heating as possible while installing an off-grid renewable system. You may use solar power to provide lights and to recharge electronic devices like tablets, computers, and cellphones.

The most practical energy source is solar since it uses photovoltaic panels to help turn solar energy into electricity. This system can be used practically anywhere in the world and just needs a few things to function. The sun is the only thing you need in addition to solar panels. However, a solar system could not be a dependable power source in other seasons, such as winter. Additionally, it performs flawlessly throughout the daytime hours of sufficient illumination. To store the energy you can use at night and on overcast days, you must rely on a bank of batteries.

The main component of several off-grid systems that use direct current produced by solar energy is battery storage. Sun-generated energy can be stored and used as needed with the help of a battery bank. The longevity of your batteries will depend on their quality; good batteries can last up to five years. Lead-acid batteries, which frequently power solar systems, have extensive charging cycles. Liquid acid batteries need to be handled carefully to prevent a disruption in your power supply.

Because they are liquid-free, absorbent glass mat (AGM) batteries are pricey but simple to maintain. Batteries can be used to store power generated by multiple-generation methods in addition to utilising solar energy. Batteries can be charged using the energy generated by generators or wind turbines. Make every effort to keep your batteries in excellent condition if you want them to live longer.

Wind Energy

Another efficient source of electricity is wind energy. A turbine is made up of turbines that are propelled by the wind, which causes the internal generator to spin and generate power. After that, the system receives this electricity, or it can be stored in battery banks. In general, wind generation systems are less expensive to put up than solar systems.

The difficulty with this energy source, though, is that some locations don't get much wind. Wind turbines could not work well where you live if there are a lot of tall trees and other physical obstacles. Another consideration is that you might not be able to provide a constant and uninterrupted power supply if your turbine is located in a region that is extremely unstable but windy. For a wind turbine to operate steadily, it must be placed in a big, open area.

Micro-Hydro Energy

Since motion is used to spin turbines to generate power, micro-hydro energy creation is not all that dissimilar from wind energy. The motion that turns the turbines in this instance is caused by water flow. The electricity generated by this type of power production is constant as long as there is genuinely enough water flowing. The only issue, though, is the high expense of investing in this form of energy production. A single household might not be able to use it.

Generators

In fact, those who live off the grid can use generators to supply them with electricity. For instance, a portable generator can be a suitable energy source if you are not living off the grid permanently. To meet your energy needs if you live in a motorhome, you might need a portable generator. Large standby generators can also provide electricity to houses that are not wired into the main grid. On the other hand, using generators for off-grid power might be expensive over time. This is the main justification for thinking about the aforementioned renewable energy sources.

What Does Living Off the Grid Means?

The concept of "living off the grid" and getting away from it all has become very popular. They want to get away from the city's traffic, noise, crowds, pollution, and technological diversions. Living off the grid actually offers a certain level of independence. Although to some individuals this manner of life may seem exotic, it truly isn't. Henry David Thoreau published Walden, one of literature's most uplifting works, in 1845. He spoke about the sense of liberation he experienced and how leaving the contemporary world behind allowed him to enjoy life to the fullest in his book.

Some people could see this way of life through rose-colored glasses because it allows you to live simply and get away from society, like in the film Into the Wild. Or they perceive suicide as an act of resistance, a means of fleeing the oppressive and corrupt system. Even though this might seem excessive, many people have this opinion about the off-the-grid way of life. While for the majority it is the earth and the environment, survival is a concern for some. Whatever the motivations of those who chose this lifestyle, it has gained popularity over the past several years regardless of what those motivations may be.

What Does Living off the Grid Mean?

No Technology

It is similar to Henry David Thoreau's way of life in the 1800s to decide to live without technology. Although we are aware that there was not the same level of technology as now, you get the idea. This is more of a primitive way of life where you hunt for food, produce your own food, and rely on your own water and gas sources. In addition, your house won't have any appliances like a refrigerator or an oven. Most people who follow this way of life are either retired or self-employed and don't hold regular 9 to 5 work. Since it is fairly tough, as one could expect from the name, it is also referred to as a "roughing it" lifestyle. To be able to live this manner, you must have strong resolve and devotion to your ideals. Although this way of living may seem excessive, some people find happiness in it. However, because you would be giving up so much if you choose this lifestyle, you must give it careful consideration.

Some Technology

Everyone may not enjoy extremely harsh play, but there are still some acceptable levels. While yet connected to the modern world, you can live off the grid. If you run your home on solar or wind energy, you'll have access to electricity for things like watching TV and surfing the internet. You can work a typical job and own a car to get to the grocery shop. It is a much more convenient and easy way of life. You are no longer dependent on utilities and are able to live the life you've always desired without making many sacrifices.

When deciding to live off the grid, there is no right or wrong answer. It all comes down to individual preferences. Living fully off the grid and leading a "primitive" life is also an option. You can live off the grid and still be linked to the outside world. whatever makes you feel content and happy.

Now that you are aware of what it means to live off the grid and the variety of options available, you may be considering whether or not this is a wise decision. It has advantages and downsides, just like anything else, and understanding them better can help you make a decision.

The Advantages of Living off the Grid

It Is Good for the Environment

Living off the grid has many real benefits, but one of the largest is that it's better for the environment. You have probably seen and heard a lot of information on TV and other media platforms about the state of the environment, about global warming, and how it impacts the globe. The problem of global warming is quite serious. Let's say we wait to implement significant modifications. In that situation, our immediate future and that of the next generations may not be as promising. Utilizing renewable energy sources can significantly alter the levels of carbon emissions. Since employing renewable energy sources like solar electricity is a big part of living off the grid, this way of life can help slow global warming. Living off the grid can have a significant impact even on just one person. Utilizing clean energy can help you reduce your carbon footprint because utilizing fossil fuels and nonrenewable energy both contribute to the release of carbon into the atmosphere.

Saving Money

Gas, water, and electricity are not inexpensive. The typical household spends over $2,000 a year on utilities, but going off the grid and relying only on renewable energy will save you money because you will never have to pay utility bills again. No need to pay for utilities, little to no eating expenses, go out to pricey devices, or make the opulent purchases you once did. Your lifestyle will be simpler and more affordable, which will have a positive effect on the environment.

Living Anywhere

Living off the grid has the benefit of allowing you to live anywhere. Simply close your eyes, visualize where you want to spend the rest of your life, then move there and construct a home. Having said that, there are a few factors you should take into account before deciding where to put your new house. If you're unsure if you need a permit, you should do some online research or contact local homebuilders. Alternatively, if only tribes exist there, you must speak with the tribe's chief. To make sure everything is done correctly and legally, some advance planning may be necessary, but it will be worth it in the end.

Away from noisy neighbors and congested cities, you can lead the peaceful and tranquil life you have always imagined.

Self-Reliance

You would be surprised at how dependent and dependant you are on so many things if you took a closer look at your daily activities. You can actually purchase food at the grocery store or order takeout, and you may cook it with any of the numerous gadgets you already own. Since you will be growing your own food, hunting for meat, and producing your own energy, living off the grid will teach you to be independent. Many people believe that developing self-reliance is a fundamental skill that everyone should learn in case of resource depletion or social collapse. You might think this is a bit far-fetched, but after everything that happened in 2020, we can genuinely claim that anything is possible.

The Disadvantages of Living off the Grid

Initial Costs

The phrase "in the long run" is important since it will show you how much money off-grid living can genuinely save you. This is due to the cost of solar panels, especially ones that will power the entire property, which is high. Despite this, many people consider these upfront expenses to be investments. This is such that using solar power is still significantly less expensive than using the domestic grid. You will also save a ton of money because you won't have to pay utility bills again, as we just explained. Additionally, as more and more individuals adopt green lifestyles or live off-grid, the demand for solar panels has grown. This indicates that there is now a demand for it and that there is rivalry between companies. What does all of this mean to you? Competition results in varying costs and more affordable goods.

Since you are constructing a new home, you will also need to pay for construction, equipment, and supplies in addition to the cost of solar energy. The price of a water system and land is another factor. Therefore, before making this significant decision, you should balance the financial benefits and drawbacks and ensure that you have a suitable budget to launch your new life.

Setting Up and Maintaining

Along with being expensive, setting up and maintaining your new solar system can be difficult. This necessitates very careful installation, planning, designing, and research. This is different from living in a typical home where everything is set up and connected for you. The maintenance phase follows the design and installation phases. To make sure your solar system is operating properly, you must monitor it. Additionally, you'll need to frequently clean your solar arrays and recharge the batteries. A professional should be used for yearly maintenance as well. Nevertheless, many modern models come with little to no maintenance requirements.

Small power

You must be mindful of your energy usage if you utilize your own solar electricity. Since you now have restricted power, you cannot use as many appliances as you once could. You might not be allowed to use an air conditioner, you'll have to switch off the lights when you leave a room, and you'll have to throw away some of your gadgets. This will undoubtedly change how you live, but given that your entire life is changing anyhow, will this present an issue for you? Before making any alterations to your life, you should carefully evaluate this.

Loneliness

People are by nature sociable animals. We are accustomed to coexisting with other people in a social setting. Off-grid living can occasionally be isolating. It's possible to go days, weeks, or even months without speaking to someone. Your mental health might not be very well served by this. Even the most reclusive people occasionally need to engage with others. Are you equipped to handle this kind of solitude? Think about that carefully. The good news is that you can live off the grid and yet engage with others. Off-grid villages exist where you can live the lifestyle you've always wanted without sacrificing contact with other people.

Consider the advantages and disadvantages before making this important choice, and decide if this way of life is right for you. If you determine it is, you should probably look for a place to fit your new lifestyle. Actually, there are a lot of nations that are more accepting of an off-the-grid lifestyle.

Off the Grid Living

By going off the grid, you can solely rely on solar energy. This idea is now widely available throughout the world thanks to cost effectiveness and a green lifestyle.

People have a lot of questions when considering going off the grid. These questions mostly relate to the necessities and advantages. Here are the answers to all of those queries:

How Much Energy Do I Need?

This is a legitimate query, however the response will depend on your individual energy requirements. A medium-sized home needs roughly 900kWh of electricity each month, however placing solar panels strategically can lower this need.

Make a List of the Most Energy-Consuming Appliances

First, create a list that includes all the appliances consuming maximum energy on the grid. These can be:

- Air conditioner
- Lighting

- Refrigerator
- Electric furnace
- Clothes dryer
- Dishwasher
- Oven and others

All of these appliances' energy sources can be replaced thanks to modern solar panels, but it would be a good idea to get rid of any that are not crucial to your home. This leads to the following point.

Sorting Movable Loads into Different Groups "Off-Grid"

The energy needed to run the refrigerator, air conditioner, water heater, and other appliances can be obtained, but this will mostly depend on the amount of energy that can be produced locally. The capacity of the solar panels, the installation space that is accessible, and the amount of sunlight received year-round are what matter most in this situation. You will be able to determine which electric loads are more necessary to transfer off the grid once you are aware of the overall quantity of energy you are capable of producing.

List any smaller electrical devices you own.

Consider the general smaller loads in your location after assessing your significant loads. Small lights, laptops, routers, and other devices can be among these loads. Even though they appear to be low energy consumers individually, a comprehensive assessment is necessary.

Switch to Energy Star certified appliances.

Designing your home for solar power would be a prudent move before going off the grid. This implies that you may redesign your household's energy consumption by selecting energy-efficient appliances. Where changes are possible, adjust the heating system, lighting, and other components.

To achieve the highest level of energy efficiency, you can even consider insulating your home as much as you can.

Can My Excess Energy Flow Back to the Grid If I Want?

Your location and utility provider determine it. In some places, net metering is accessible. If your solar system is connected to a grid, this method enables you to credit your extra energy back into the grid.

You must confirm whether net metering is permitted in your state because certain states do not permit it, which prevents you from choosing to send extra energy back to the grid.

Lower Energy Costs Each Month

The biggest advantage of adopting off-grid power is certainly the cost savings on energy. Although there are some initial costs involved, the installation costs only need to be paid once, saving you money on subsequent monthly expenses.

Installing solar panels can result in monthly savings of up to $100, depending on your location, home size, energy usage, etc. In the majority of major cities, high-quality panels will often last more than 20 years, saving you over $20,000.

In addition, you can employ net metering if it's available in your state. You must continue to be connected to the electric grid in order to sell any extra energy to the grid and make monthly savings. Many people think that using net metering is preferable to being completely off the grid. This will allow the entire facility to be energy efficient, which will increase your monthly savings.

With net metering, you can typically "sell" your extra energy to the grid in the form of energy credits. You won't have to pay for grid electricity when you need it (due to insufficient solar or increased energy needs). The use of net metering as opposed to being completely off the grid has many advantages.

Prevention of Blackouts

In many places, blackouts are frequent occurrences. An off-grid power source can be useful if you reside in such a place. Your lights continue to be on, assuring the safety and security of your business. The lights will receive electricity from the solar power system, enabling you to run them continually as you like.

You can unwind about electricity availability even if blackouts are not an issue. Grid electricity won't matter.

Ability to Be Used for a Range of Purposes

Solar energy is more advantageous for everyone due to the range of energy sources. Solar energy produced by solar panels can be transformed into heat or electricity. Just the correct system will do. As a result, you can operate a variety of equipment in your home or place of work without having to worry about running out of power.

Low Maintenance Is Needed

It's commonly known that solar systems require little upkeep. Once or twice a year only requires basic cleaning. Naturally, the quality of the first installation will determine this. Additionally, solar panel cleaning is an affordable option. Panels from reliable manufacturers come with guarantees, enabling you to put off paying maintenance fees for a long time. Your solar panel manufacturer will cover all expenses, no matter what happens.

The effectiveness of the panels is governed by their maintenance. As cables transport energy from one component to another, you should concentrate on having your cables fixed on a regular basis. When compared to the savings from going off the grid, this won't be too pricey.

Continuous Improvement

All big firms are investing in the creation of more advanced solar technology because it is the way that energy will be consumed in the future. Thus, if you move to solar energy, you can benefit from the most recent developments in power generating. You have the highest chance of using power and energy if you use the most modern technologies.

You can modify your power system as technology develops to achieve more efficacy and efficiency. Grid energy used traditionally does not offer this possibility.

Now that you know, switching to solar energy offers some clear benefits. Making your final selection will be made easier if you are aware of these benefits. Pay special attention to chances for cost-savings and living quality. You can always have your energy needs met while saving money.

Now that you've learnt everything there is to know about solar energy and solar panels, get ready. If desired, choose the most appropriate panels, reliable manufacturers, and utility sources. Make your daily power consumption more economical and efficient by using what you've learnt. Start by calculating the precise cost of the energy needed and the amount needed. After that, move on to finding and installing solar panels and other components as the next stage.

Things to Consider Before Mounting a Solar Panel

- Consider the amount of power you need before deciding on an off-grid structure.
- What is the required battery storage for your solar system?
- Estimate the number of solar panels that can produce the required voltage.
- Choose a solar charge controller.
- Purchase an inverter for your solar system.
- There should be a balance of systems (BOS).

Surveying Your Site

In conjunction with the National Renewable Energy Laboratory (NREL), you will also decide on the output and size. Of course, using it is free.

The location of your property and a number of other details are entered into the PVWatts calculator to determine how much electricity you can generate with a specific solar system size. The PVWatts, however, cannot assist you in measuring your azimuth, the area where you wish to install your solar, or the slope of your roof. These will need to be measured by you.

Slope, Area, and Azimuth Measuring

The measurements that characterize the location and positioning of your future solar panel are these three numbers. Additionally, you'll require these numbers in order to use the PVWatts calculator.

Slope

The tilt angle of the solar panel is determined by the slope and angle of your roof.

The relationship between the vertical rise and horizontal run of a particular roof plane is known as a roof slope. Therefore, if you have heard someone refer to a roof slope as (3:12), it signifies that the roof plane rises three inches over a horizontal run of twelve inches that is level. It is crucial to measure the slope of your roof with the utmost safety and precision.

To measure the roof slope, you will need the following tools:

- A ladder
- Level
- Tape measure
- Pencil

Here's how to calculate the roof slope using a right-angle triangle's geometry:

1. Using a ladder, you must first ascend to the roof of your home.
2. You will need to fabricate a flat surface, such as a flat board, that spans over the uneven portions of your roof if it has uneven planes (i.e., there is no level pan from which to take measurements).
3. Mark a 12-inch point on the level using your pencil. (You don't need to worry about this if your level is already twelve inches. Only those with a higher rank are eligible.
4. Then, starting from the beginning of the flat surface you made on the roof, incline it using your level.
5. Next, using your tape, measure the distance from the level's twelve-inch mark to the roof surface.
6. Your slope is whatever ratio you obtain. (For instance, if your tape measure registers three inches, the slope of your roof plane is 3:12).

7. The rise-run slope you get should be converted to degrees. For instance, PVWatts (Tap on the 'i' in the 'tilt (deg)' row on the'system info' page) offers conversion tables online. Additionally, a scientific calculator is available. Calculating the inverse tangent of the rise/run ratio will give you the degree angle. Simply divide the rising value by the run value and press the calculator's inverse tangent button.

You may use a level app on your smartphone to position your solar panel on the ground. It could be a component of the 'compass app.'

Area

Solar panel installation can be done for free on your roof or outside.

Use a very long tape measure that is at least 25 feet long to obtain quick and precise measurements. If you don't have a longer tape measure, you can mark the roof with some chalk at each measurement instead. It may be annoying and stressful, but it will accomplish the task.

The steps listed below can help you determine how much space is available on your rooftop:

1. Make a quick sketch of a simple area view map of any roof regions that are appropriate for solar panels before scaling your rooftop to obtain measurements.

2. Note on the map where chimneys, roof vents, satellite dishes, and plumbing pipes are located in relation to other impediments and penetration points.

3. You can now ascend to the rooftop following this.

4. With your tape measure, measure the length and width of each roof plane on your map. An acceptable area might be at least 25 feet long and 20 feet wide, measured from the ridge of the roof. It is also possible to install fewer solar panels in more concentrated sections of the roof.

5. Subtract more than five to ten inches from each ridge and roof edge relative to the installation area you measured. To leave a small space gap around the array, subtract is

required.

6. Next, take precise measurements of each obstacle and point of penetration and record them.

7. While taking measurements, leave a few-inch buffer on the sides of each obstruction.

8. You might have to subtract the shaded area from your total if a rooftop obstruction prevents any installation area from being used during the hottest part of the day.

Before finalizing your plans for the layout of your roof, check with your local authorities.

You must measure the installation area for ground mount arrays on your land or yard.

Azimuth

This is related to the specific directory that your roof or ground mount array faces. For instance, the degrees of west, south, north-west, etc.

Azimuth measurement methods:

1. The azimuth of a solar array pointing north is 0 degrees.

2. The azimuth of a solar array pointing directly east is 90 degrees.

3. A solar array facing directly south will have an azimuth of 128 degrees.

4. A solar array facing straight west will have an azimuth of 270 degrees.

5. An old-fashioned compass or a compass app on a smartphone can both be used to calculate azimuth.

Smartphone use is much simpler. Why? You don't need to take magnetic declination into consideration because your smartphone calibrates itself to your location.

To Use a Compass App:

1. Start your smartphone's compass app by doing nothing.

2. When given calibrating instructions, follow them.

3. Once the compass app is configured, stand with your back to the roof where the array will be placed.

4. Next, place the phone directly in front of you while holding it flat in your hand.

5. The exact angle that your roof plane faces is provided by the compass program. The azimuth of your roof is that angle.

6. The traditional compass can be used in the same manner. Just be sure to take your particular location's magnetic declination into account.

7. It is optimal if the array faces due south (an azimuth of 128 degrees) if you intend to put your solar panels on the ground.

Sizing Your System with PVWatts

Here is how to use the PVWatts calculator to determine how much electricity you can generate annually:

1. Go to the PVWatts calculator online and spend a few minutes there. This calculator is available at https://www.pvwatts.nrel.gov.

2. Select "Go" after entering your home address, city, or state.

3. Select the map position nearest to your home on the following page, or choose your city's default location.

4. After that, press the large orange "Go to system info" arrow.
5. Click the large orange arrow marked "Go to PVWatts result" on the following page.

6. At the top, a number is shown in huge font. The figure represents how much AC electricity your site is capable of producing annually.

Easy, huh? There aren't any unneeded commercials, no log-in registration requirements, and, most importantly, it's free! The calculator already has the system size and other standard inputs.

PVWatts appears to be quite straightforward, but there is much more you can do with it to increase the accuracy of your results. For more precise results, input your unique information on the 'System Info' page. Six fundamental design criteria are provided on the PVWatts 'System Info' page to help you size your PV solar system. Let's quickly go over the features. Click the 'i' button to the right of each parameter to view these adjustments.

System Size in D.C.

The system size is the total solar array's rated D.C power output. PVWatts's standard system size is 4 kW. You might use a solar panel array made up of 20 PV panels, each rated at 250 watts. You must multiply 20 by 250 to get the system size for the entire panel array (20 x 250 = 5000 watts = 5 kW). The default system size in this instance is (4 kW), but you can choose to adjust it to reflect the actual size of your calculated solar panel array.

Module kind

For more precise results, you can choose which of the three types of P.V. solar panels you use using the module type parameter. Modules made of crystalline silicon (c-So) or thin-film technology are also options.

Array kind

Here, you can choose whether to use a rooftop mount, ground mount, or another sort of solar panel attachment. Simply choose it, whichever one you are using.

System Failures

This is the sum of all minor efficiency declines brought on by imperfect equipment performance and actual operating circumstances. Due to these circumstances, the quantity of A.C. power transmitted into your home cannot be equaled by the real D.C. power received from the sun.

The tilt of your photovoltaic solar panels is measured in degrees.

Azimuth

After making all of these adjustments, go to the "System Info" tab and click the orange arrow that says "Go to PVWatts results." The figure you receive is the typical annual kilowatt-hour production you can anticipate with a (5 K.W.) system after inputting more precise results. You might receive a different answer from a (5 K.W.) system (unless you entered a different system size).

Snow

Snow-covered solar panels don't generate much electricity. When snow falls, the sky is cloudy; this is enough to generate power.

Things are made more difficult by a solar panel that is covered in snow. Use a wood-handled brush or mop that won't scratch the solar panels' glass if you want to speed up the removal of snow from them.

The solar panel gets a lot of power from the sun again once the snow is removed or melts away and the sun shines again.

Mistakes You Should Avoid When Installing an Off-Grid Solar Power

Insufficient Solar System Size

The most frequent error is sizing a solar power system wrong. Usually, this is brought on by mental shortcuts and basing power requirements on variables that do not precisely reflect the amount of power utilized. For instance, you might think you use X amount of power based on your most recent utility bill but in fact you consume a different amount depending on the season. When selecting how much power to use, other aspects such shifting climate temperatures, panel location, and efficiency must be taken into account.

Using Your System for Rent

Solar power systems are only a wise investment if you own the system. The leasing of a solar system has a lot of drawbacks. The region offers tax incentives for putting solar systems in residences, but if you leased your system, you would not benefit from them (they would go toward the actual owner). Second, even while renting could seem more reasonable on a monthly basis, you will pay too much interest. Additionally, it would force you to commitments that are in direct opposition to your independent way of life. Therefore, I advise you to purchase your system and grant yourself the satisfaction of having it.

Not Making Plans

Many people make the error of estimating how much electricity they currently use when deciding to create an off-grid solar power system instead of taking into account how it will be utilized in future homes.

Buying Affordable Solar Panels In an effort to save money

I can well understand why a cheap solar panel would be appealing to a buyer (I enjoy a good deal myself). But you need to be able to tell the difference between a sale and a cheap solar panel being placed on the market. There have been some new players as a result of the rising demand for solar energy, with many of them supplying systems produced in China at outrageous costs.

Missing the Warranty Restrictions

In the solar sector, the length of a product's warranty is believed to be closely tied to the manufacturer's faith in its product. In other words, it is thought that a product is of poor quality and performs poorly when the guarantee duration is limited. Keep this in mind when searching for solar items and focus on those that have the longest lifespans. Recognize the conditions of the warranty and the services it covers. Please do not be fooled by performance warranties that claim a product will last for the duration of the market. A manufacturer's warranty should be found and have a ten-year lifespan.

Use of Sun-Tracking Systems

Many people think that installing sun-tracking equipment will enable them to enhance the quantity of energy their solar panel array produces. On the other side, investing in a sun tracking device might add to an already long list of expenses. This is due to the fact that purchasing a tracker (which may or may not help "track" more sunshine) is preferable to acquiring more solar panels and being proactive about your need for increased energy production. Don't anticipate sun-tracking devices to last as long as your solar panels because they are prone to system breakdowns.

Refusing to Purchase a Battery Monitor

Battery monitors are useful tools that offer important details about the condition and well-being of the battery bank. By identifying issues with the off-grid solar power system or current power availability, this device helps with bank maintenance. For people who are still learning about solar energy and need the most support in comprehending their solar power system, this material is extremely helpful. Additionally, the battery monitor will record and quantify the total amp hours used by the system, enabling you to keep an eye on a household's energy consumption.

Not Making the Upgrade to Energy Efficient Appliances

When your off-grid solar power system is completely operational, the last thing you want to happen is for energy to be wasted or used improperly. When you use gadgets that require a lot of needless energy, you waste energy. You may conserve as much energy as you can while still enjoying the benefits of utilizing electronics and appliances by altering your lifestyle choices and choosing more energy-efficient gadgets. A small change in lifestyle, such as using LED lightbulbs instead of incandescent ones, can save a lot of energy.

Refusing to buy a backup generator

When establishing an off-grid solar power system, generators could seem like an unnecessary cost. Nevertheless, they are a crucial part of the system, particularly if you have made the decision to live completely off the grid. Being able to power their homes however they see fit is one of the main reasons people opt to live off-grid. Unpredictable weather circumstances are beyond your control, but you can prepare for them. In the event of a calamity, which could happen at any time, your generator will be useful. You cannot plan an off-grid solar power system under the premise that the climate will always be stable and neutral.

Inability to adopt new technology

Since the solar power sector is still relatively young, new updates and products are constantly being produced. This is wonderful news for you because there are always opportunities to increase the effectiveness of the solar power system. The increased demand for solar goods also implies that owning a solar system is becoming more affordable. By staying current with new technology, you may make more efficient component purchases and build a solar power system that is more effective overall. Additionally, it will make you aware of less harmful alternatives to installing a solar power system, lowering your overall carbon footprint.

Living On the Grid vs. Living Off the Grid

Pressures and reliance are the main distinctions between on the grid and off the grid. However, I'm not here to mislead you. I'm not a bad salesperson. I want to let you know. To the best of my abilities, I want you to have access to the details contrasting these two ways of existence. I would like to apologise in advance if I come out as prejudiced toward one of these lifestyle choices, but that is where I am at the moment.

Using only grid power

Sometimes it can feel like life is perfect off the grid. For the top ten percent of earners, it undoubtedly feels perfect all the time, but even that, I doubt. The others go to work, most likely work nine to five, and you return home.

Life starts to take on a mechanical and mindless quality. You complete chores and generate income. When you go out to eat as much as you can because you are so busy, money is spent.

Your home receives utilities, and money is expended. You keep an eye out for the newest consumer goods that appear annually around Christmas.

In essence, this is a textbook example of economies of scale. Because you are so skilled at what you do—as opposed to those who are better at creating things, preparing meals, or delivering utilities—people turn to you for your services. While it does save some time, the quality and significance are sacrificed.

What did you take away from that meal?

What does this day's purchase represent to you? What will happen next year?

What happens next when the electricity is lost? Waiting for someone to attend to it is required. It can take several hours. What would happen if an entire town lost power? Before you receive your service, you might have to wait while all the other repairs are completed. What I mean when I say that the Amish handle things off the grid is this. Instead of waiting for others, they rebuilt entire villages with their own sweat and work. Before the lights were turned back on in the Big Apple, they resumed their daily lives.

What transpires next if you are not happy with something? Your product malfunctions or the service is subpar someplace. Do you call the company's customer service and try to speak with someone there? Not if you want to keep your hair on your head.

The mechanical component is here. Everything seems distant and monotonous.

Living Off the Grid

Living off the grid, on the other hand, is taking responsibility for yourself without doing it in a monotonous or onerous manner. It is not so much about being unconnected. If something really isn't working, it's your fault. It's not as bad as it seems because guess who can fix it. You may! You don't have to go without electricity to survive. You may, but that is entirely up to you. You can generate your own electricity if you still desire it. Living off the grid does not require you to disregard hundreds of years of progress in humankind. Instead, it can be remotely produced without a connection to the power grid, to be perfectly frank.

Because more people are pumping their own well water into their homes, running water is increasingly noticeable. Given that the US government adds fluoride to public water systems, it should come as no surprise that many individuals oppose this forced water modification. The abundance of fresh water sources nearby is startling, but we'll get to that later.

At this point, you just need to realize that everything can be done independently without relying on utilities or the additional tax that goes along with them. You don't have to rely on others in real life either. If a difficulty arises, you are the answer.

Realizing that you are the issue and that you can only address your own issues while unable to address those of others greatly reduces your level of frustration.

But aside from utilities, you can accomplish practically everything on your own. In a reasonably sized personal garden, you can grow your own fruits and vegetables. Additionally, it's easier than you might think to manage to raise a small herd of livestock. Additionally, you don't truly need to be concerned about industrial farming's usage of pesticides or how food is handled during processing.

You are essentially producing off the land at this stage. Your shelter is integrated into the land, which also provides utilities and supports local agriculture and animal life. But there's still more you can do. You can start creating your own set of crafting talents. Making and building things yourself.

At this point, you could be thinking, "He's just telling me you can do this without really giving me information." I appreciate your focus.

History of Solar Energy

Although it has become a very important part of our efforts to produce and use renewable energy today, this wasn't always the case. The long history of photovoltaics demonstrates how the idea of solar power has changed over time to become what it is now. Solar energy has been less expensive over time, particularly in recent decades. This is due to improvements in the design and underlying technologies of PV systems. 15 years back. To understand where silicon solar technology is today, let's quickly review the development of solar energy.

Solar Energy in its Early Years

According to theory, man has been using solar energy since the seventh century B.C. But even at this early stage, the forms were relatively simple. A magnifying glass was used by numerous prehistoric societies to focus solar energy on light tires. For example, the Romans and the Greeks lit torches for religious purposes using these solar reflectors. Chinese civilization has recorded the existence of these blazing mirrors. When the Romans attacked Syracuse and forced a withdrawal, Archimedes, a Greek scientist, is said to have used the reflecting qualities of a bronze shield to focus sunlight on the wooden Roman ships and set fire to them.

Sunrooms, which are still widely used today, were another early application of solar energy. These are spaces with large windows that are intended to direct sunshine into a space. These south-facing windows are thought to have been present in Roman bathhouses, effectively converting them into specially heated sunrooms.

It is thought that some native American tribes, such as the Anasazi tribe, constructed south-facing homes on cliffs later, about the year 1200 AD, which were useful for retaining the warmth of the sun during the chilly winter months.

The 16th and 17th centuries saw the rise in popularity of more advanced systems. Horace de Saussure, a Swiss physicist, created the first solar oven. Even though solar panels weren't created until a while later, this significant event represented a significant shift in how solar energy was used.

A History of Solar Panels

It is crucial to remember that numerous scientists have to contribute throughout time in order for solar panel technology to advance to its current state. As a result, there have been disagreements and ambiguities over the creation date of solar panels as well as who should receive credit for the invention.

Edmond Becquerel is widely regarded as the solar cell's creator. The photovoltaic effect, which explains how energy generation can be boosted when two metallic electrodes are placed on a conducting solution, was initially discovered by the French scientist. Later on, the invention of selenium-powered PV cells owed a lot to this 1869 finding.

Willoughby Smith later discovered selenium's photoconductive properties in 1873. William Grylls Adams and Richard Evans Day used this discovery as a springboard to discover that selenium had the ability to generate electricity when exposed to sunshine. A few years later, Charles Fritts, who is frequently credited as the creator of solar cells, created the first solar cells manufactured with selenium wafers in 1883.

Three people—Gerald Pearson, Calvin Fuller, and Daryl Chapin—are commonly credited with creating the solar cell as we know it today. PV cells are now made of silicon rather than selenium. We were able to harvest the sun's rays and utilize them to power our equipment for several hours each day for the first time in living memory thanks to those three scientists' invention of the PV cell in 1954 at the Bells Laboratory. However, the conversion rate of sunlight by these early solar cells was only 4%, making them not very effective. Just 25% of what a contemporary PV cell is capable of producing.

Initially, satellites were the main objects powered by solar panels in space. The Vanguard I satellite's radios were powered by a solitary one-watt solar panel in 1958. Other satellites including the Sputnik-3, Explorer, and Vanguard II also utilised PV solar cells, and in 1964 NASA's Nimbus spacecraft was powered by a set of solar arrays with a combined output of 470 watts.

In 1973, the first solar-powered home was constructed. In 1973, the University of Delaware debuted the Solar One building. It was the first instance of an integrated photovoltaic system for a building that utilized a combination of solar PV and solar thermal energy. Instead of using solar panels, this structure had them built right into the roofing.
However, the first solar panels were incredibly inefficient, making it difficult for them to be adopted broadly right once. From 1957 through 1960, significant advancements in solar conversion efficiency were made. Hoffman Electronics produced a number of breakthroughs on the efficiency of PV cells at this time, which was a significant turning point for the technology. The innovation saw a record leap in solar cell efficiency, from 8% to as high as 14%. Later, in 1985, the University of South Wales succeeded in producing solar cells with a 20% efficiency.

Higher conversion efficiency has been made feasible by further advancements in PV cell design throughout time. As an illustration, studies conducted at the National Renewable Energy Laboratory in 1999 produced solar cells with up to 33.3% efficiency. Later, in 2016, researchers from the University of Wales broke this record by achieving a 34.5% efficiency. It is crucial to remember that these were merely laboratory experiments and not mass-market productions.

The cost of solar energy has decreased over time as a result of the greater efficiency. The cost of PV solar cells has decreased dramatically during the last few decades. Consumer demand has increased as a result, fueling the industry's growth. In the US, more than a million installations had been completed as of 2015.

Up to $300/watt used to be charged for solar panels. By 1976, this price had fallen to only a little bit over $100/watt. Recently, the price of installing a solar panel has dropped to as little as $0.50 per watt. Solar energy is becoming more and more well-liked as a dependable energy source as a result of cost reductions and other advantages.

What is Solar Power Energy?

Finding sustainable energy has generated a lot of controversy throughout the years. Scientists have been searching for a cost-effective and efficient answer ever since the energy crisis of the 1970s. The first solar panel was developed in 1954, thirty years earlier. Even while it worked, there were still a lot of issues to iron out, particularly with regard to its scale, effectiveness, and global accessibility.

But the price of producing and distributing solar panels has dropped significantly since the 1970s. Many families all over the world can now afford solar panels. You may build your own mobile solar devices using simple instructions, and it has become one of the fuel sources with the fastest growth in the globe.

How do solar cells function?

The most potent instance of fusion in our solar systems is the sun. The results are visible by glancing out a window. Scientists had figured out that it creates a potent agent here on Earth as early as 1839. The discovery that some materials created energy when exposed to light was made by a French scientist named Edmund Becquerel (Knier, 2008).

Despite being a scientific advance, this wasn't really considered an energy source. Albert Einstein didn't actually develop a theorem for photovoltaic energy—the capacity to produce a current when two materials were exposed to light—until 1905.
Solar technology advanced alongside scientific knowledge. After it was founded in 1954, NASA started to take solar panels' utility seriously. As a simple and affordable replacement for conventional fossil fuels, solar panels were taken into consideration for the first space flights. Since there were no other fuels in space, they only used solar energy. It was a scientific advance.

Solar panels have improved and are now widely accessible. The shrinking of solar cells is what causes this true miracle. Due to the smaller battery being used, solar panels are more effective at transmitting the current they generate.

Photons, the tiny units of light, are used by the battery to capture the energy of the sun. A nuclear reaction causes the transmission of these particles. Photons typically produce heat when they reach Earth. By going for a walk outside during the day rather than at night, it is simple to observe the impacts.

A semiconductor substance that responds to light makes up the battery. Electrons bonded off of the substance when photons struck it, starting an electrical reaction that produces energy. The current received transmits energy in the form of direct current when electrons strike either side of the positive and negative ends of the battery. In essence, the energy generated by the sun's reaction with the semiconductor material enters the current directly.

In What Ways Do Solar Panels Produce Energy?

It is already known that solar energy is produced when protons strike the semiconductor battery in a solar panel, causing electrons to scatter and form a current. That is still a long way from supplying energy to a machine, though. In spite of the fact that solar panels generate electricity, they rarely do so in a way that is suitable for our everyday equipment.

Due to the way the electricity is transmitted, alternate current (AC) and direct current (DC) are not the same. Electrons move through a current directly in direct current. A "charge" period is necessary for alternating current, which is the most widely utilized current in the world. Because it can be incorporated into generators and other machinery so readily, AC is quite common.

Due to modern electronics' reliance on DC, AC may appear to be going extinct, yet it is still extensively used in autos and city grid systems. This makes using an inverter to convert DC power into AC for solar cells necessary.

PV Inverters

Consider an inverter to be the organization's top manager. It functions to provide a current that other systems can use. It functions throughout the remainder of the system, storing energy as required.

For many years, central inverters were the rage. This is due to the fact that, for general inverter use, they were the only sort of technology accessible. For all the work that each solar panel in a system performs, they serve as a basic inverter. In other words, if a system has 10 solar panels, the central inverter's responsibility is to transform all of the DC energy into AC energy. The inverter is unable to generate an output large enough to support the system's cells when it grows unusually large. If there are numerous solar panels as well, the price increases dramatically.

Over the past few years, the price drop for solar cells has gotten increasingly pronounced. Micro-inverters are to blame for this. Power is distributed more evenly with micro-inverters. Imagine being the only employee in a company with 500 employees. The manager is capable of handling every aspect of the job, but he or she is unable to visit every employee and impart all of the knowledge at once. Imagine a different company has 500 employees, but they are now separated into five-person teams. A sub-boss oversees and assists all of the staff. Put that to use with solar cells now. The idea of a central solar inverter system looks absurd if there are thousands of solar cells. The smaller solar cells in the solar panels communicate with each other through the micro-inverters. As a result, they can use more solar panels to create electricity more effectively.

Racking

Even though it could seem to be among the least significant aspects of solar energy, mounting a solar panel is crucial. Solar panels are frequently tipped to the side when they are installed on buildings or other structures. This is not a coincidence. When solar panels are exposed to direct sunshine, they perform at their best. Directly upward-facing solar panels will not receive as much sunlight as ones that are slightly angled.

The only location on Earth where the sun will pass directly overhead is at the equator. The seasons change everything else on Earth, too. If you live in the northern hemisphere, the sun will always be pointing south; if you live in the southern hemisphere, the sun will always be pointing north. As a result, solar panels that face entirely upward may not always receive direct sunlight.

A lot of people decide to install their solar panels on rotatable platforms. They will receive more electricity from their solar panels thanks to this method of ensuring longer time in direct sunlight.

Additionally, the racking shields the wire from the elements. The electronics must all be kept in a safe environment because solar panels are meant to endure for 20 to 25 years. Solar panels themselves have been upgraded to be more waterproof, but the rest of their components still require adequate protection from the elements.

Charging Control

Charge controllers regulate the amount of electricity that solar panels produce. The charge controller, despite appearing to be an unneeded component of a solar panel configuration, is essential for maintaining solar energy regulation. With the help of charge controllers, batteries are kept from being overloaded by ensuring that the energy from solar panels is constant. The intermediaries in the solar energy system are charge controllers. They keep a constant current whether transferring the energy straight to batteries or to a residential electrical system.

Consider a system lacking a charge controller. The solar panels take in the greatest energy from the day's energy gain. They transmit powerful energy to a power grid as a result. You might notice powerful lightbulbs blazing. The lights may flicker as the sun moves and stops shining directly on the solar panels, but they get dimmer with time. Similarly, if a cloud moves into the path of the sun, the energy collected from the reduced amount of light will cause the light to dim.

Any extra energy delivered to the charge controller is controlled appropriately, resulting in a steady stream of light. There wouldn't be any energy variations. Therefore, utilize a charge controller as you would with any configuration.

Batteries

Very little can be done with the energy collected if there are no solar energy storage facilities. The charge controller regulates the electricity that the sun provides before being stored in batteries. Batteries are the most expensive component of the design, aside from solar panels, so take caution while selecting batteries. Ensure that your batteries may be fully charged and discharged. Battery life should be checked every five years and replaced as necessary. If they are worn out, think about replacing them.

What Number of Solar Panels Are Required?

How many solar panels you need for your project will rely greatly on the available area, how much energy will be used, and how many people will be using the power. For instance, more solar panels are required to power a large home in order to make up for the enormous amount of area, whereas a small home may only require half as many. Solar panels will be needed in significant numbers for homes that truly utilize a lot of energy for appliances. However, you might just use a few if you're traveling and solely use power to jump-start the refrigerator. Similar to how a family of 10 will need more solar panels than an individual.

Analyze your power bill to determine the amount of energy you will use. On an electrical statement, the power provider also includes the number of kilowatts utilized, even though many customers just look at the amount owed. To start, look at the average amount of kilowatts consumed annually. The wattage listed on the container should be taken into account when choosing the next lightbulb. If there are 50 lights in your home and they are all left on for two hours each day at a power consumption of 50 watts, there are 5000 kilowatts used daily.

Next, figure out how many sunny days there are on average each year. There are probably fewer sunny days in New York than in Arizona if you reside there. The amount of electricity that solar panels receive greatly depends on the weather. Solar power is harmed by cloudy days. Solar energy might not be the best option if you live somewhere where it rains almost daily.
Based on the kilowatt-hours of electricity you consume, inquire with many businesses about the number of solar panels you'll need. By receiving many opinions, you can save more money, and smaller businesses will typically offer you a better bargain. After a year, if you still need more solar panels, add more.

However, allow the experts to direct you to the ideal number for you.
It is conceivable to power the entire Earth with solar energy because the sun generates a virtually limitless supply of it. While it may be easier said than done, it is nevertheless possible to harness a significant quantity of energy for personal use from solar power.

It is getting easier and easier to harness solar energy because micro-inverters transform it into AC current for mobile gear. All you need to connect things is a little bit of knowledge. The installation of a functional solar power station requires the use of charge controllers, racks, and batteries; therefore, it is important to consider the cost and effectiveness of each component.

Remember that your environment has a significant impact on the amount of solar energy you receive when estimating your future energy requirements. Latitude also has a significant impact on a solar panel's efficiency, in addition to weather. Think of someone who resides in Miami,

Florida as opposed to Nome, Alaska. Florida would receive more energy annually than Alaska if both had the same number of sunny days. Solar efficiency is greatly influenced by shorter days and less direct sunshine.

The Science Behind Solar Energy

Learning how solar energy functions and how it is converted into the electricity we often use in our homes is the best place to start when it comes to truly utilizing the full potential of solar energy.

Solar energy is the heat and light that the sun radiates onto the surface of the earth. The majority of people believe that solar energy exclusively consists of solar heat. The sun also produces light, or "sunshine," and the more of this energy the sun emits, the more solar energy we typically receive on Earth. The amount of radiation that actually reaches the surface of the planet varies depending on location.

Solar thermal energy, solar photovoltaic energy, solar design, and solar heating are some of the technologies that assist in capturing this energy. The majority of the electrical equipment in our homes are powered by these technologies, which convert the heat and light from the sun into electrical energy.

Photovoltaics (PV) and concentrating solar-thermal power (CSP) are the two most widely used solar energy technologies.

Photovoltaic solar technology

These are the principal components and machinery for converting light into electricity. To this end, they often incorporate PV cells, each of which is capable of producing one to two watts of power. The cells are occasionally sandwiched between safeguards due to how strong the light may be.

The cells may occasionally be linked together in a chain to create bigger components known as panels or modules in order to improve conduction or the ability output of the cells. One by one, or in groups, these modules are joined to form arrays. An whole PV system is created by connecting one or more of these modules to the electrical grid. These systems can be adjusted to meet almost any power requirement, no matter how big or little.

However, the PV system as a whole includes both arrays and modules.

The systems also include mounting frameworks that make it easier for the modules to be exposed to the sun and other components that collect the (DC) electricity generated by the modules and transform it into the (AC) electricity that we typically use to power the appliances in our homes.

The Golden State solar PV behemoth, which generates 579 megawatts of electricity, has the largest PV system currently known to man.

Power Systems Using Solar Thermal

To deliver the high temperatures required to generate energy, these devices gather and concentrate light. Reflectors and receivers are the two primary parts of the energy collectors used in the systems. Light is captured and focused onto the receiver by reflectors (mirrors), which makes it easier for energy to be collected and transferred to the turbines.

The majority of systems use a heat transfer fluid, which is typically heated before being pumped through the receiver to produce steam. This steam is then converted back into energy in an extremely rotating engine, which drives a generator that generates electricity.

To ensure that the light is directed onto the receiver throughout the day as the sun changes positions, trailing plans are built into solar thermal power systems.

These systems also contain a thermal energy storage technology that makes it possible to store energy during the day and use it later in the day or at night.

Solar energy towers, linear concentrating systems made up of linear Fresnal reflectors and parabolic troughs, and solar dish/engine systems are the three basic types of concentrating solar energy systems that are sometimes used to classify solar thermal power plants.

Systems for Linear Concentration

The rectangular, long, curved, or U-shaped mirrors used in linear concentrating systems are intended to capture solar energy. These mirrors help concentrate the sun's rays on the receivers that cross the mirrors.

As a result of the heated fluid heated by the concentrated light flowing through the tubes, it is then transported to a device where it boils in a typical turbine generator to produce energy.

There are two primary types of linear concentrator systems: parabolic trough systems, where the receivers are positioned at the center of the parabolic reflector, and linear Fresnel reflector systems, where one receiver tube is positioned on top of numerous other reflectors.

A linear concentrating collector station may have numerous collectors arranged in parallel rows, usually facing north and south for the best absorption of solar radiation.

1: Fresnel reflectors that are linear

Similar to parabolic trough systems, these systems focus light onto a receiver that is affixed to the top of the mirrors. Due to the Fresnel lens's close focus and large aperture, this is frequently possible. The sun's energy is concentrated by these Fresnel reflectors to around thirty times its usual intensity.

Additionally, there is a specific type of Fresnel reflector known as a compact linear Fresnel reflector that has several absorbers inside the mirrors. These many receivers enable the mirrors to change their angle in order to lessen daylight obstruction from neighboring mirrors, reducing the amount of value and material required.

2: Troughs in parabolas

On the other hand, parabolic troughs have reflectors that direct sunlight onto a pipe receiver positioned in the center of the conic troughs. As the sun moves from east to west, this collector revolves with it.

Due to the troughs' parabolic shape, the sun's intensity is concentrated from thirty to one hundred times higher than usual, with the ability to produce temperatures of more than 750° F.

Many of the solar thermal power plants in the world with the longest operational lifespans use these parabolic trough technologies. One of these facilities is in the Mohave Desert in Golden State and is one of the largest solar thermal power plants in the world, with a net summer electrical capacity of around 365 MW.

Engines / Solar dishes

In that the container catches light, these solar collector systems function very similarly to the satellite dish you use for reception. Makers generally match smaller flat mirrors that have been fashioned into a dish, driven by the need to make the plate more affordable and practical.

The Stirling engine, which uses the fluid heated by the receiver to shift the pistons and generate mechanical power, is the most used solar collector. The generator or generators that produce electricity are then powered by this mechanical energy.

These solar dishes direct the sun's light directly at the center of the plate. Solar dishes provide significantly more energy than linear concentrating systems with working fluid temperatures above 1380° F.

In order to maximize the process of energy collecting, the gadget needed to supply the electricity is positioned in the center of the dish.

Towers for solar power

The solar energy tower system concentrates light onto a receiver at the top of a tower by using several heliostats, flat-shaped mirrors. The sun's heat and light can be concentrated up to 1500 times more intensely with these technologies.

Because water has the potential to move energy from one area to another through heat conduction, some towers use it. In some systems, liquid nitrate salt is also used because of its excellent heat transport and energy storage properties. These systems can produce power at night or during overcast weather since they have the potential to store thermal energy.

You can locate those solar energy system types most commonly used in the world.

The Core Benefits of Solar Power Energy

There are various energy sources available today, but solar energy often outperforms the others due to its many benefits.

Here are a few of the most popular justifications for why you should give solar energy in your home serious consideration.

It emits less pollutants

Solar energy doesn't release any hazardous or greenhouse gases into the atmosphere, unlike other energy sources like fossil fuels, which are bad for the ecology. As a result, it contributes to maintaining a safer and cleaner environment compared to conventional energy sources, which release about 23 metric tonnes of greenhouse gases into the atmosphere per year.

These gases are the main contributor to global warming and climate change and are the main cause of the numerous metabolic disorders we are currently facing.

It is regenerative indefinitely

A limitless natural resource is the sun. Because it can be collected from every location on the planet without running out, solar energy is actually one of the most dependable and sustainable sources of energy.

You are guaranteed an endless supply of solar electricity every day because the sun rises every day. This is especially true when compared to fossil fuel, whose supply is steadily declining and will eventually become exceedingly unreliable.

Because solar power doesn't require a live connection to the grid in order to create electricity, it's an excellent option for distant locations.

Anywhere on Earth may easily harness solar energy, which can supply your home with the necessary power. This can improve people's quality of life, particularly in Sub-Saharan Africa.

Maintains foreign energy spending

Your typical daily energy costs can be reduced thanks to solar power, which can also cut the price of oil used to generate electricity.

The US alone spends $13 million each hour on oil from the country in the Persian Gulf. This is a significant sum of money, especially when you consider how much money is required to refine this oil and supply the electricity from it.

When it comes to solar energy, all you need is the appropriate hardware, and since the sun is a free natural resource, you won't have to spend any extra money gathering the raw ingredients.

It's generated new employment possibilities

Researchers and inventors have created various devices to this end since solar energy was discovered. Additionally, when new technologies are developed, a growing number of jobs in production, research, sales, and installation are created.

Additionally, businesses are switching to solar energy as a substitute, which means solar energy is giving more people jobs. Now that solar power generation is a subject taught in schools, it is a technical talent that can open up job opportunities for many people.

It is affordable

Aside from the expense of purchasing and installing the solar equipment, using solar power is almost cost-free after the system is up because you won't be need to continue paying monthly energy bills in order to continue receiving its benefits.

This enables you to accumulate the cash you would have otherwise spent daily on electricity. There are occasions when you may even produce enough energy to sell to others for a modest fee.

It is trustworthy

Solar energy is particularly dependable since, unlike other alternative sources of energy, it is not susceptible to utility company failure, economic uncertainty, terrorism, or even burnouts from being utilized too much.

You won't have any restrictions on using solar power, and you won't have to be concerned about price increases brought on by a variety of variables. Solar energy is consistent, and while its effectiveness may gradually decline over time due to wear and tear, you can still rely on it to meet your demands.

Solar energy can be used for extended periods of time, even in cloudy or dark conditions, because it can be stored in a battery.

Offers generous tax advantages

The tax incentives associated with solar energy use are among the most frequently discussed advantages.

Today, if you qualify as a homeowner and decide to install solar panels, you may be able to receive a federal tax credit equal to 26% of your total upfront costs; a good example is New York, where you may receive a $5,000 tax credit.

With the aid of these incentives, you may accelerate the return on your investment in the solar energy sector.

Increases the value of your property

The cost value of your property is certain to rise because solar power reduces so many home maintenance costs. Because of this, residences with solar electricity tend to cost more than those without it. This is because a home without solar energy will be far more dependent on the grid, which is more unpredictable than a home that generates its own electricity.

Additionally, installing solar electricity is a terrific investment that, if done correctly, might yield a significant investment return in the future.

Low cost of upkeep

As long as the system is properly configured, solar energy effectively generates itself with little oversight. Additionally, as new technology advances, solar panel maintenance gets simpler and simpler.

With warranties built in, new solar technologies are less expensive to maintain in the event of an issue. You won't have to pay for unneeded repairs or new solar power producing equipment as long as you genuinely maintain your panels clean and your connection is adequately built. Your solar power system won't need any additional expenses to operate at its best.

Off-Grid Solar System Overview

Wiring

When putting up their system, whether it is their PV system at home or on their boat, or even their hi-fi audio, the majority of people assume that choosing the proper cable is a simple matter. They are unaware that wiring is one of the most popular techniques to increase efficiency and build a better PV system overall. The incorrect wiring can endanger your appliance's safety or worse, decrease the performance of your system.

Having a diagram of your setup and identifying the location of your battery bank is a smart place to start. The size of the solar panels, the length of the cables connecting the solar array to the solar charge controller, and the cabling leading from that to the batteries must all be considered. Cables are also required to connect the batteries in a parallel configuration. Then, connect cables to run from the batteries to the inverter after measuring the cables for DC power to the outlets you anticipate having DC power. Again, wire is required to connect that location to the outlets. Calculating the peak and average loads on each wire is the best technique to ensure that the wiring is done correctly. Therefore, for example, the cabling that connects the inverter to the outlet that powers the 15,000 BTU air conditioner needs to be thicker (lower gauge) so that the appropriate amps can be delivered to the unit. Make the most of the lengths to the outlets you require since wires might be pricey.

Batteries

Now that the panel system's construction is complete, we can make the entire thing work properly. Making ensuring that it runs flawlessly both during the day and at night is one way to deal with this. We depend on non-traditional energy sources. Another strategy is to reserve a lot of energy during the day that you may use at night. As a result, we require an energy storage system to handle the extra electricity that our solar panels produce.

Everyone is aware that batteries are a practical way to accomplish this. When we talk about people who are having financial problems because they bought batteries but don't understand the benefits of this technique, things become more complicated.

Some people might think that the energy these solar panels produce is insufficient to be stored. Since not all of the items we have linked to the solar panels are active during the day, this doesn't seem to be the case. You might be too busy elsewhere or there might not be enough people in all the rooms, for instance. The selection of the proper batteries becomes another problem. Batteries come in two different varieties. The first to appear were the liquid batteries. When there was no alternative technology to compare them to, these batteries were frequently used to store vast amounts of power. They were a remarkable discovery at the time.

Science and technology have made great strides, and the energy markets are now flooded with dry batteries, which don't require any liquid to function.
These batteries use mechanisms that are different from those found in current batteries. They can be utilized for a lot longer than traditional batteries and have a much bigger storage capacity. Because they are smaller, lighter, and safer, they are actually gaining in popularity. They do have a drawback, though, in that they are somewhat pricey. If you have a lot of money to spend, you could use this battery. If money is scarce, you might be able to get by with the dry

ones. Since not all liquid batteries are made equal, it is critical to select the best options on the market that are within your budget.

Charging Control

Charge controllers regulate the amount of electricity that solar panels produce. The charge controller, despite appearing to be an unneeded component of a solar panel configuration, is essential for maintaining solar energy regulation. With the help of charge controllers, batteries are kept from being overloaded by ensuring that the energy from solar panels is constant. The intermediaries in the solar energy system are charge controllers. They keep a constant current whether transferring the energy straight to batteries or to a residential electrical system.

Consider a system lacking a charge controller. The solar panels take in the greatest energy from the day's energy gain. They transmit powerful energy to a power grid as a result. You might notice powerful lightbulbs blazing. The lights may flicker as the sun moves and stops shining directly on the solar panels, but they get dimmer with time. Similarly, if a cloud moves into the path of the sun, the energy collected from the reduced amount of light will cause the light to dim.
Any extra energy delivered to the charge controller is controlled appropriately, resulting in a steady stream of light. There wouldn't be any energy variations. Therefore, utilize a charge controller as you would with any configuration.

Inverters

Consider an inverter to be the organization's top manager. It functions to provide a current that other systems can use. It functions throughout the remainder of the system, storing energy as required.

For many years, central inverters were the rage. This is due to the fact that, for general inverter use, they were the only sort of technology accessible. For all the work that each solar panel in a system performs, they serve as a basic inverter. In other words, if a system has 10 solar panels, the central inverter's responsibility is to transform all of the DC energy into AC energy. The inverter is unable to generate an output large enough to support the system's cells when it grows unusually large. If there are numerous solar panels as well, the price increases dramatically.

Over the past few years, the price drop for solar cells has gotten increasingly pronounced. Micro-inverters are to blame for this. Power is distributed more evenly with micro-inverters. Imagine being the only employee in a company with 500 employees. The manager is capable of handling every aspect of the job, but he or she is unable to visit every employee and impart all of the knowledge at once. Imagine a different company has 500 employees, but they are now separated into five-person teams. A sub-boss oversees and assists all of the staff. Put that to use with solar cells now. The idea of a central solar inverter system looks absurd if there are thousands of solar cells. The smaller solar cells in the solar panels communicate with each other through the micro-inverters. As a result, they can use more solar panels to create electricity more effectively.

Grounding

When electrical charges from your PV system are discharged into the ground or the earth, this is known as grounding or earthling. It serves as protection whilst working with anything electrical.

The Distinction Between Earthling and Grounding

When grounding your solar system, you would use a conductor that could transport current, whereas when grounding the earth, you would use a conductor that could not carry current.

To ensure the safety of both the user and the equipment, grounding or earthing your solar system is primarily done to transfer the potentially dangerous roaming charges surrounding your off-grid photovoltaic system to the ground. Although your system already has fuses and disconnects, it may seem superfluous to ground or earth your solar equipment.
You should still ground some pieces of equipment, though. For instance, grounding your solar panel makes sense because it generates a lot of electricity. You can choose to ground any area of your solar system, though.

Off-Grid Solar Panels

Solar cells are a great investment because they provide the energy for all photovoltaic (PV) systems. These devices are made of silicon, which is the most economically attractive and widely available semi-conductive material for use in the manufacture of solar panels. The same electrons, neutrons, and protons that make up every other element in the periodic table also make up silicon. Albert Einstein discovered the physics phenomena known as the photoelectric effect, which is the basis for the way solar cells generate electricity. The physics underpinning it may be simply and clearly explained as involving the conversion of light energy into electrical energy. There are many different wavelengths included in solar radiation. Based on this spectrum, ultraviolet and visible light, or heat and light, can be distinguished as the two main components of solar energy.

The wavelength range of solar energy that solar panels may use to produce power is found in the visible light zone. The photons that make up sunlight have kinetic energy that causes them to drift in the direction of the Earth. When these solar-derived particles get to the silicon interface of the solar cell, they transfer their kinetic energy to the electrons of silicon atoms. The silicon becomes conductive due to this energy transfer, permitting a small electric current to flow. Without getting too technical, a photovoltaic module is a structure that is created by connecting solar cells in series and parallel to combine their output. We may increase the electric current and voltage values, as well as the power outputs, that are practical for use in common market applications, thanks to this. All solar panels fall within the aforementioned concept.

In reality, there are a number of differences amongst PV modules. As a result, we may classify solar panels according to the technology that powers them. Solar panels of these three varieties include rigid (monocrystalline), semi-flexible (polycrystalline), and flexible (amorphous).

There are many different shapes and sizes of panels.

All of the information up to this point is relevant to solar panels in general. There are certain variations among PV modules that are noteworthy. These modules have the highest light-to-electric energy conversion efficiency available on the market (for recent leading brands), with conversion efficiencies ranging from 19 to 22%. As a result, they are being researched for use in a variety of PV applications, particularly those with a restricted amount of space for the installation of solar cells. Due to their better performance, they are more expensive. Because they make the most use of the available space to produce the most electrical energy, these PV modules are also in high demand in the market for recreational vehicles (RVs), cabins, and boats. Due to their beautiful black or dark blue color, they are very frequently used in a variety of applications that are related to the house. Monocrystalline solar cells' outside edge has a rounded form due to the manufacturing process.

The Difference Between Off-Grid and On-Grid Solar Energy

It's time to consider your options now that you are familiar with solar power and its advantages and disadvantages. Even though it's what most people prefer, just because you're living off the grid doesn't mean your solar power system has to be completely disconnected from the grid. Some people who live off the grid, however, are okay with being connected to the electrical grid. Although the idea behind solar energy is unchanged, the actual system configuration may change. It can be on-grid, connected to your area's local energy grid, or off-grid, a standalone power system. Hybrid systems are also available.

On-Grid

When it comes to solar energy, on-grid solar power systems may be the most often used configuration. Both houses and businesses utilize this type of system to reduce utility costs and transition to a more ecologically friendly source of energy. On-grid solar systems don't require batteries to store the generated electricity, in contrast to stand-alone power systems (SAPs). They substitute solar inverters instead. Additionally, because on-grid solar systems are connected to the local power grid, any extra solar power generated by your home is exported to the nearby grid in exchange for payment.

Contrary to popular opinion, unlike hybrid power systems, on-grid solar power systems do not function during blackouts for safety reasons. It would be extremely unsafe for a solar inverter to continue providing electricity into the damaged grid because a blackout is frequently the result of damage to the electrical system. Because power would still be flowing through the grid, this would put everyone who is repairing it at risk. However, with hybrid networks, the solar power system can be automatically disconnected from the rest of the grid so you can keep supplying electricity to your home. An on-grid power system can use batteries, but there are a number of factors to take into account.

In an on-grid arrangement, any extra energy passes via a meter at the switchboard that determines how much power you're buying or exporting. Different countries will have different systems in place for this process. In some locations, the amount of electricity your meter registers as your export to the grid. In other nations, the meter may track the full solar energy output from your system before it gets to the switchboard. In any case, local residents use the extra electricity your solar power system exports. On the other side, you will need to import electricity from the public grid and pay for it if your usage is more than your production.

This technique works well in areas where there may not be enough sunlight throughout the year, which means your off-grid home's solar power generation may not be adequate. To make up the deficit in those circumstances, you can obtain electricity from the nearby grid. Large residences that consume excessive amounts of electricity without relying on the local grid are an excellent candidate for an on-grid solar power solution because of the high expenses. In this way, solar energy can be used to meet the majority of your energy needs, and when it is insufficient, power from the local grid can be used to meet the remaining needs.

On-grid systems have a few other benefits besides lowering utility costs. It's quite simple to install them. A company can save a lot of money on specific utility bills with this configuration without having to invest a ton of money up front. Additionally, with on-grid configurations, you can generate money by exporting your excess power to the nearby grid whether you're a household or commercial user.

Off-Grid

Off-grid solar power systems require batteries to store the electricity generated by the solar array because they are not connected to the public grid like on-grid systems are. Off-grid systems have one drawback: careful planning is required to ensure that you can supply your home with enough electricity throughout the year, even when the sun is not at its highest point. Additionally, your batteries must have adequate storage space to hold all the energy generated. Off-grid solutions' primary drawback is that they are significantly more expensive than on-grid alternatives. The high price of batteries and inverters is mostly to blame for this.

Although this method is expensive, it is the best option for people who want to live off the grid and are independent of the public power grids. For someone who lives in a distant area and doesn't even have access to the local grid, which is true for many individuals, this is also their only alternative. Although remote areas can pay to extend electricity lines from the main grid, doing so would be extremely expensive and almost certainly more expensive than installing an off-grid solar power system. According to US Department of Energy estimates, it will cost between $15,000 and $50,000 per mile to build power lines from the main grid to rural regions, which would be enormous sums for someone living off the grid in a remote area.

Solar panels, a charge controller, batteries, a grid box, an inverter, the rest of the system's components, and the mounting framework make up the fundamental setup of an off-grid system. There are many off-grid system types with slightly varied component arrangements, but we'll get to those later. This system is perfect for rural areas since it saves solar energy throughout the day so that it may be used at night. Blackouts won't effect you because you're not connected to the main grid, so whatever happens to it won't affect your ability to get electricity.

Your home appliances suck up the electricity produced by the solar panels in an off-grid solar system. After your needs for electricity have been met, any extra energy is transmitted to a battery bank. When all the batteries are charged up, the charging stops. You use the power stored in the batteries to run your appliances at night because the solar power system isn't functioning due to the lack of sunlight. For those periods of the year when you don't have much electricity stored due to clouds, experts advise purchasing a backup generator. In those circumstances, a generator will be necessary for both your power needs and battery charging. Since there is no public electricity grid to supply you with power here, unlike in on-grid systems, you should absolutely approach the generator as a necessary rather than optional part of your off-grid power arrangement.

Hybrid

By utilizing the advantages of each, hybrid systems let you choose between the two alternatives that are ideal for you. As with an off-grid system, the main power configuration in this case involves installing a battery bank to store extra solar power. The main distinction is that you would also have access to the main grid. The grid acts as a backup, supplying your home with the necessary quantity of electricity on cloudy days or when your battery supply isn't sufficient to meet your needs. Hybrid systems are intelligent enough to charge your batteries between midnight and six in the morning, when electricity is most affordable, during off-peak hours. In the long term, this helps you save money on your utility expenses.

The creation of a hybrid solar power system can be done in one of two methods. The first is the one where a battery bank is used, as we just explained. In accordance with the time of day, the strength of the sun, and your consumption, this will be refilled either using solar energy or electricity from the main grid. For a typical home, there is an average usage, but during hotter or warmer months when you might utilize electrical heating appliances, this consumption can soar. As soon as your batteries are fully charged, the second strategy is to enable your system to export any excess power to the main grid.

Which System Is Best for You?

You may start choosing the off-the-grid solar power solution that would be most suitable for you now that you are aware of your alternatives. Off-grid electricity will be the main subject of this book because it is what most people choose because it offers the greatest degree of independence and ensures that you won't be dependent on anyone for your energy demands. Additionally, it functions best for people who live in distant areas without access to a local power grid. However, if all three are offered, you'll need to do some math to determine which is best for you.

Actually, there are a few things you need to think about before choosing an off-grid system for your house. Obviously, the most important determining element is cost. You must determine which of those systems is the cheapest over the long run after considering your loads. Don't focus only on the initial expenditure because, regardless of how large the original cost is, you can break even in a few years. The permits should also be taken into account. To learn about the rules governing solar electricity, check your state, county, and/or local legislation.

Solar panel installation must be done by a licensed expert in some states, such as California, and cannot be done at home. You must have a licensed expert install the solar panels in California, even if you live off the grid with no connection to the main grid, or you risk fines. The laws can potentially be more severe in other US states.

States with severe laws governing the types of materials you can use can be anticipated. Others won't even permit you to do so unless your house is a certain size before you can install a solar array system. In order to be sure you are not breaking any laws, you should actually check with the county office in your area or at the very least the internet. Hiring a certified solar panel professional who is familiar with the many types of permissions required as well as any other important information would be an easier approach to go around this.

On-grid or hybrid solar power systems will involve gaining authorization and approvals, so you should be aware of the necessary permits and documentation if you choose to use such. Exactly asking an electrician to connect you to the electrical grid is not something you can do. So, be sure to learn about the required approvals early on to avoid having to put a stop to the construction of your off-the-grid farmhouse because of bureaucracy.

Types of Off Grids Solar Power Systems

Solar power systems that use direct current (DC)

The form of power that is most frequently generated from real sources, including solar cells and batteries, is known as direct current (DC) power. The term "direct current" (DC) refers to the movement of electrons in a single direction.

The flow of electrical charge, or, to use another name, electrons, in circuits that employ direct current (DC) is unidirectional; unlike AC Current, which periodically changes its flow direction.

The waveform most frequently connected with alternating energy is a pure sine wave, as shown in the picture below.
DC (red curve), direct current. Time is actually measured along the horizontal axis; current or voltage along the vertical.

What typical DC uses are there?

Examples of the types of sources that are most frequently utilized to produce this kind of electricity include thermocouples, solar cells, and batteries.

DC power is frequently utilized in low voltage, low current applications such as battery charging, automotive, aerospace, and other low voltage, low current applications. Every solar panel created today generates DC electricity. Two examples of typical applications for DC power in the photovoltaic industry are portable solar power systems and various kinds of off-grid products. By doing away with the necessity of a solar inverter, which actually changes direct current into alternating current, these kinds of systems may have their prices cut.

1. Firstly, let's talk about generation. One example of a sustainable and clean energy source that can be used to generate DC power is solar energy. As a result, it has no impact on pollution of any type. The majority of the processes involved in the creation of alternating current (AC) power supplies, such as thermal energy, nuclear energy, etc., are tied to intricate designs and are therefore vulnerable to pollution.

2. A DC supply has a very high energy efficiency because the magnitude and polarity are always the same. The magnitude of the AC power supply varies over time, which indicates that it is less energy efficient than DC. Additionally, there is a power factor issue, which contributes to noise.

3. A DC supply is ideally suited for usage in electronic circuits since it is continuous, which is one of its characteristics. As a result, a source of DC power is required for the operation of all electrical devices.

4. Motors and other items powered by a DC supply, such as lights, are significantly more energy-efficient than items powered by an AC source.

5. DC circuits are particularly efficient since there is no noise produced by them and very little power is lost due to inductance or capacitance effects on the DC supply.

6. Another important advantage of employing DC (Direct Current) is energy storage. Alternating current, often known as AC, cannot be stored; however, direct current, or DC, may be exceptionally easily stored in a battery or super capacitor.

7. Distributing and transmitting DC power is substantially less expensive than doing the same with AC power.

8. Only DC power sources, not AC power sources, are used by all modern digital electronic communication systems, control systems, and technical developments.

Direct Current or DC drawbacks

1. Alternating current (AC) generators are simpler than direct current (DC) generators. In practice, direct current can never be produced at a very high voltage due to the commutation problem.

2. DC power cannot be transferred over a long distance since it cannot be stepped up or down, despite the fact that it is less sophisticated and more affordable to transmit and distribute than AC power.

3. The procedure of going from AC to DC is rather simple, but going the other way, from DC to AC, is not only more challenging but also inaccurate.
4. Even while alternating current (AC) machines use less energy than direct current (DC) machines, the procedures involved in their design and operation are more intricate. An AC motor's design is significantly simpler than that of a DC motor, for example, if we compare the two types of motors.

5. Important concepts like electromagnetic induction and shunt reactance cannot be applied in the real world because the inductance and capacitance effects do not exist with a DC source.

6. Electrical machinery and gadgets that use DC supplies, like DC motors and DC circuit breakers, are more expensive than those that use AC supplies because of the more complex nature of their manufacture.

7. It is not feasible to directly increase or decrease the DC voltage. Direct current (DC) requires a more involved technique to step up or step down the voltage than alternating current (AC), which may be accomplished with simply a transformer. Prior to increasing or decreasing the voltage, the direct current (DC) must first be converted into alternating current (AC). Every time the DC voltage is increased or decreased, there is a higher loss of power.

Coupled AC Systems

A battery-based inverter/charger receives power from an existing power source in your system through AC coupling, which in turn charges the battery bank. You merely add a battery-based inverter/charger and a battery bank to your already-existing system rather than changing its configuration. The battery bank just needs to be big enough to run your vital equipment during a power loss, such as the good pump, the refrigerator, the electrical parts of the heating system, a few lights, and the phone charger.

How does it function?

An additional electrical sub-panel known as a critical loads panel can be connected to by an inverter or charger. The connections for the most important circuits can be found in the circuit breaker box. Additionally, a connection has been established between the grid and the inverter. When the inverter detects a grid issue, it shuts off all grid connections and only powers the devices that are hardwired into the panel for the crucial loads. Thus, it complies with the requirement that it not transmit electricity down the wires while yet allowing you to have some power in your home.

If you have a solar system that uses AC coupling, it will be connected to the panel for the critical loads if you already have a grid-tie inverter. The inverter, which is used when the grid is operating, distributes electricity to the rest of the house. Any extra energy is then fed back into the grid, where it is used in the same way as before to flip the meter around.

However, if there is a grid issue, the transfer switch will switch to the opposite position, cutting off all connections to the grid and connecting to the panel for the important loads. The device immediately recognizes the output coming from the inverter as being that of the grid because the grid-tie inverter is also connected there.

The grid-tie inverter won't shut off if there is enough daylight. By sending solar electricity to an inverter that also serves as a charger, it keeps the battery bank charged. As a result, you are using the battery bank to supply power for your most critical loads while the solar panels recharge the batteries during the day. When the power is restored to the grid, the transfer switch is reset, the grid-tie inverter continues powering the entire house, and the meter is turned the other way.

Since the AC charging is fed back into itself via the AC output, it does not use the three-stage charger that is housed in the AC input. Every inverter and charger manufacturer has a technique for avoiding overcharging batteries.

Advantages

- a higher efficiency (up to 97% efficient) when utilized to power AC loads all day long, like air conditioning, pool pumps, and hot water systems.
- For larger systems with capacities more than 6 kW, there are lower overall installation costs.
- ability to use several solar string inverters in different places (AC connected micro-grids)
- On-string solar inverters with a capacity of 3 kW or higher typically have dual MPPT inputs. This enables the deployment of long strings of panels in various orientations and tilt degrees.

- Advanced AC linked systems can make use of both AC and DC coupling, however some "controlled" lithium battery types do not allow for this.

Disadvantages

- a reduction in the total efficiency of the charging mechanism, which varies between 85% and 90%.

- The size of the inverter charger (i.e., Victron 1:1 ratio) may place a cap on the solar inverter's capability.

- For less extensive power generation, high-quality solar inverters could be a bit pricy.

Hybrid Solar Systems

Direct current (DC) coupled solar battery systems that are grid-connected are one way to think of hybrid energy systems. They actually come in a wide range of designs and typically use a hybrid or multi-mode inverter. The high voltage MPPT controller or controllers and the battery inverter/chargers are incorporated into a single housing unit in today's hybrid inverters. However, higher voltage (400V+) battery solutions have proliferated recently. Initially, 48V lead-acid or lithium battery systems were compatible with hybrid inverter technology.

Batteries: high voltage vs. low voltage

The current generation of "high voltage" batteries have an operational range of between 200 and 500 V DC (with a nominal value of 400 V), in contrast to the more prevalent 48V battery systems. The advantages of this include increased effectiveness and decreased cable losses. This is possible because the solar panel typically operates at a voltage between 300 and 600 volts, which is quite similar to the battery voltage. Higher voltage batteries (400V) and hybrid inverters that work with these batteries use lithium battery systems that run between 120 and 500V as opposed to 48V.

Two different battery arrangements are possible for higher voltage batteries:

1. The solar panel array and inverter are DC-coupled.
2. A compatible hybrid inverter is DC linked directly to it (as indicated below).

Since the bulk of solar arrays operate at high voltages, commonly between 300 and 600 volts, high voltage batteries use DC-DC converters that are incredibly efficient and have very low losses.

Advantages

- Low price, compact size, and simple assembly are all necessary
- High-efficiency battery charging (around 96%) Battery substitutes that are both small and modular
- an improvement in efficiency and a reduction in cable size when using high voltage batteries
- 'Some' existing solar arrays can be retrofitted with the new technology.
- Hybrid inverters are increasingly more widely available on the market.

Disadvantages

- Some systems can't function as a backup power source.
- Many systems with a backup experience delays during a blackout that last between three and five seconds.
- In general, off-grid sites are not a good fit for these systems because transformerless hybrid inverters have a low surge rating and no generator controls.

Backup Battery Systems

In fact, AC batteries, often referred to as AC-linked batteries, are a relatively recent innovation in the field of battery storage for homes that are wired into the electrical grid. These batteries enable rapid AC battery coupling to a brand-new or pre-existing solar system. Lithium battery modules, an inverter/charger, and a battery management system (BMS) are the components that make up AC batteries. These components are all assembled into a single, straightforward, and compact device. Most dwellings can easily be connected to this unit.

Because the (transformer-less) inverters are often not powerful enough to run a home totally off-grid and are unable to withstand the surge loads of a range of appliances, these systems are typically only intended for use in grid-connected homes and not off-grid homes. dwellings that are wired into the electrical grid are referred to as grid-connected dwellings.

Advantages

- Simple to install; it may be adapted into homes with solar panels already in place.
- A technology that adds energy storage at a low cost
- The majority of the time, extensible systems with modular components are simple to deploy in new places.

Disadvantages

- 90% less efficiency due to the conversion procedure (from DC to AC and back to DC).
- A backup supply cannot be operated with certain types of AC batteries.
- The design usually does not permit off-grid installation use.

Things to Take into Consideration

For most people, the cost of installing solar power at their residence or place of business is the main issue. When compared to the typical electricity bill, solar panels and other accessories are still quite an investment, even though their costs have decreased over the past several years. In most circumstances, solar power can take anywhere from 10 to 20 years to pay for itself, even with people who use a reasonably high quantity of electricity. After that, you'll essentially be utilizing electricity for free. Many individuals are unaware that the labor costs to install a solar power system make up a sizable amount of its cost.

You could cut the cost of the solar setup in half if you performed the installation yourself because the labor cost is such a significant issue. In other words, instead of needing ten years to break even, you could achieve it in five, and by the sixth year, you would be consuming energy for free. Most individuals would find it to be a far better offer. The difficulty of installing it yourself is the drawback. The actual labor involved in moving the panels, drilling holes, and fastening them to the frames is rather simple. They are a collection of readymade parts that just require assembly. Finding a technique to place solar panels such that you would maximize production and decrease losses is difficult.

For instance, if you've ever gone camping with solar panels or installed solar panels on your car or RV to power the onboard refrigerator and other equipment, you'll know that merely having the panel on the roof does the trick. But that does not necessarily imply that you are making the most of the system. Despite having a 400 watt solar panel, you may only be using 230 or 250 watts of it. Since a vehicle or mobile setup uses less panels and it is difficult to have a properly optimal solar setup on a moving vehicle that is continuously changing directions and altitudes, this difference can be overlooked.

Installing solar panels for your home, where you'll be producing hundreds of watts of power and using this system constantly, is a whole other ballgame. More crucially, since panels put in your garden or on your roof won't move, you may be able to get up to 80% or even more efficiency out of them if you perform the job correctly. You want to be able to receive the best return on your investment given that installing a residential or business solar power system may be very expensive.

Fortunately, it is doable, and with a little knowledge of what causes a solar system to function efficiently, you can even do it yourself.

Let's examine some of the most crucial details you must actually take into account in order to make the most of your unique solar system.

Location

Many people believe that the roof is the ideal location for solar panels and that it must be where they are put. This may have been true a few years ago when we only had panels that could be installed in a certain way, but with the introduction of more recent types of panels, this idea is quickly evolving. Users can choose from a wide range of solar panels today that can be applied in a number of various ways.

For instance, there are solar panels available today that are fashioned like window glass and some of them are fully transparent, much like a genuine window. You can install these panels in place of your standard glass windows, freeing up space on your roof for additional uses. Although the fully transparent version costs a little more, there are less priced alternatives that aren't fully transparent, offer a little more privacy, and also generate power.

In a similar vein, solar panel sheets are also available. These are flexible and may be placed anywhere. They resemble a large sheet of plastic that has been tightly coiled up. These can be used extremely well for residential purposes while being very popular in mobile applications. You don't need to bother about building sturdy structures or actually placing a solar panel on your roof; all you need is a little room. A lot of lamps are available that already have solar panels attached to them. This approach avoids the inconvenience of having to install a light and then wire it up to the mainboard, where it can draw energy from above-ground solar panels. Typically, these are external lights put outside a building. These lights only require that you set them up outside and turn them on. That's all there is to it; the light charges throughout the day and turns on at night when it detects that the light of day has vanished. In view of the fact that security lights frequently remain on throughout the night, installing them in your home is a great strategy to conserve energy.

You can buy solar tiles if you want a tiled roof and don't want to destroy it to make room for some solar panels. Although these roof tiles have solar charging systems built into them, they also produce electricity. They appear exactly like conventional roof tiles. This may be the ideal solution if you currently have roof tiles placed and are thinking about upgrading. You'll actually save a ton of money and time thanks to how simple it is to install, and you'll get all the advantages a solar power system has to offer.

You are not required to use the space on solar panels if you genuinely live in a home with an attic, a wooden roof, or a roof that might not support the weight of a solar setup. Solar panels can always be put in place on the ground in the backyard or any other bare area with adequate exposure to sunlight.

Making the switch to solar-powered versions of electric geysers and air conditioners is another excellent approach to actually conserve space on your roof. The requirement for solar panels is entirely eliminated with a solar geyser. Instead, the solar pipework on the solar geysers employs a similar technology to heat water using sunlight; as a result, you have access to hot water throughout the day as well as a large supply of hot water that can be used after the sun sets. The solar panels will take up much less area thanks to this, and the procedure will be far more effective because electricity won't be required at all.

Space Measurement and Modification

The quantity of space you have available will play a significant role in figuring out what kind of solar solution will work best for your residence or place of business. You have the choice to build an entirely solar-powered home in which you won't have electricity after the sun sets, or you may install your panels in tandem with batteries so that you have electricity even after hours. Having some sort of power reserve will be required if you plan to live off the grid permanently.

You may require a lot of room for the panels themselves in addition to managing all the lighting, ventilation, and appliance systems in your house. For instance, you would require enough solar panels to provide 10Kw of power, which is often sufficient for the majority of residential homes. You might be able to fit the panels on your roof, depending on the type you select.

Solar panels with ratings of 150W, 200W, 250W, 300W, 350W, and 400W are the most popular. Some of the newest solar panels may produce up to 500W more power. For instance, you would require approximately 67 panels to produce 10 Kw from 150 W plates, whereas 20 panels would suffice if you choose 500 W panels. You are looking at approximately 540 square feet if you choose the 150W panel versus 160 square feet if you choose the 500W panel, with the typical panel measuring 2Ftx4Ft or 8 square feet per.

The fundamental benefit of using roof space is that it makes use of area that would otherwise go unused. A ground-mounted system can be just as effective if you have plenty of space and don't mind giving up a few hundred feet to install solar panels.

The fundamental goal of any solar panel installation is to ensure that all of your panels have access to adequate sunlight and are positioned optimally to maximize the amount of sunlight that is available. It can also be effective to place some of the solar panels on the roof and others on the ground.

Generally speaking, the voltage of the panel increases with the actual wattage of the panel. A panel with twice the wattage might therefore theoretically provide you with somewhat more than twice the amperage and voltage if a 150W panel were to provide 8 amps and 15 volts of power.

The greatest option for the majority of homes is the one that requires the fewest changes to the physical structure of the house and is the least expensive to install. In general, roof-mounted solutions are less expensive than ground-mounted solutions since they need less material to build the frame that the panels will rest on. However, in some circumstances the ground-mounted option can end up being less expensive. Pole-mounted options are typically the most expensive to install, but they also maximize available space. You can put pairs or sets of four panels on poles in place of installing all the panels on a large frame. People typically have a fixed amount of roof space available when hanging them. Then, in order to generate the required amount of electricity with the fewest number of panels, it becomes necessary to employ higher wattage panels.

Panels Are Adjusted for Sun and Shade

Shade is the only factor that can reduce a solar panel system's effectiveness. The problem is not cloud cover. It is the shadow produced when obstructions to the sun, such as chimneys, poles, or dishes, are present. Each panel contains a lot of modules, which is a concern. In order to produce the quantity of power the panel is rated for, a particular amount of energy must be created when light strikes the entire panel. The maximum quantity of electricity a 500W panel can produce is reduced if even a little amount of shade covers a section of it. If there are any obstacles, such trees, close to the panel, cut them to a suitable length or, even better, get rid of them totally.

Your best option is to use a hybrid solution if you have roof features like chimneys or air vents that you just cannot ignore. You can install some of your panels on the roof, some on poles, and some on the ground rather than having all of them on the roof where some of them might be shaded. This will ensure that all of your panels are operating at full capacity and help you make the most of the actual area that is available.

Panels are Adjusted for Roof Size

The alternative is to build a frame that matches the size of the roof. For instance, if your space is only 250 square feet, you'll need to choose a panel with adequate wattage to meet your requirements. The typical panel has an area of 8 square feet and can range in power from 150W to 500W. 32 solar panels on a roof with 250 square feet of usable space. The greatest power rated panels will provide 16 Kw of power, while the lowest powered panels will produce about 5 Kw. People frequently have hundreds of square feet of space on their roofs, but the issue is that only some of it faces the sun. To generate as much electricity as possible, solar panels must face the proper direction in addition to being in the sun.

Placement of Your Panels

To receive the maximum sunlight during the day, some solar installation professionals advise putting your panels south, while others advise facing them west. The sun spends the majority of the year in the southern region of the sky for households in the northern hemisphere. The panels will receive the most sunlight exposure during the daytime hours if they are pointing south. You want to face the South Pole when orienting your solar panels toward the south, not the True South or the Magnetic South. For homes in the southern hemisphere, the situation is reversed, and solar panels should face the North Pole.

Similar to this, by tilting the solar panels between 30 and 40 degrees southerly, you may establish a perpendicular angle that will cause the light to strike the panel with maximum intensity. The maximum quantity of electricity can be generated when the light strikes the panel at a perpendicular angle. The angle at which you install your solar panels can vary, and a general rule is to keep the angle of the solar panel equal to your geographic latitude; hence, in some circumstances, this angle may reach as high as 50 degrees.

Having the proper tilt angle also lessens the buildup of snow, dirt, and other material on the panels. Snow or other environmental variables can occasionally completely block light. Snow slides off more easily and dust gathers more slowly on the panels with a decent angle than it would on a flat or more inclined surface.

In certain circumstances, according to specialists, the solar panel should face southwest. This is so that sunsets in some areas, such those in northern Europe, can occur as late as 11 p.m. and dawn to be a bit later. Additionally, because the sun never sets entirely below the horizon in certain areas, residual light is visible every single day of the year. The panels benefit from more light during the late-sunset hours if they are facing southwest rather than south. Peak sunshine hours occur when the sun is directly overhead the solar panels if they are facing flat south. The additional sunlight that is available after the hottest part of the day cannot provide much energy. The modest inclination towards the west makes greater use of the later hours and can even support the creation of energy late into the evening by utilizing lingering light. However, the same idea still holds true for houses in the southern hemisphere.

When choosing the direction of the panels, some experts also take your power company's payment system into account. For instance, using south-facing panels will be more efficient if your billing system is tier-based and your electricity costs increase as you use more of it. This is because you want to create the most electricity possible overall.

The greatest electricity charge is in the evening (peak hours), when people use the most electricity, hence it would be preferable to have westward-facing panels with a time-of-use billing system. If you have westward facing panels, you can generate your own electricity during these times and won't need to use the grid.

By adding batteries to store electricity for use during the daytime when there is an excess of output, you can also lower your electricity expenses even more. To further limit how much energy you consume from the grid, you can use this additional energy at night or during times of high demand.

Essential Tools and Equipment

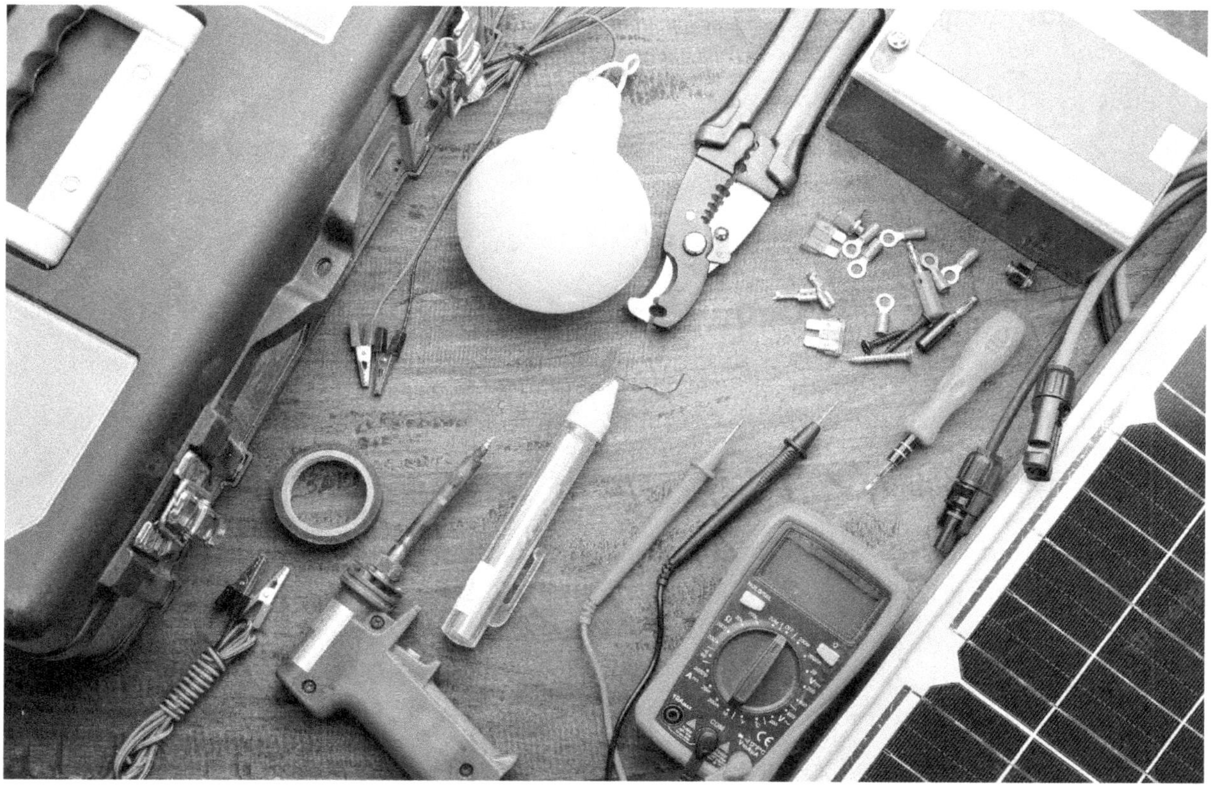

In response to the switch to solar energy, highly specialized tools for solar power installation have been developed. Others of these items are made specifically for solar power installations, even though some of them are simple to obtain around your house. These tools range from straightforward gadgets that let you cut the solar connections to quite sophisticated ones.

Learn about the fundamental tools needed to set up your PV system if you intend to install it yourself. Solar power expenses are reduced by DIY installations, but if you don't have the necessary tools and decide to hire a contractor, your upfront expenditures will go up.

You must think about your safety in addition to possessing the required installation abilities. Since you obviously want to avoid any DIY-related mishaps or injuries, you must gather all the necessary instruments before you begin your DIY installation project. In some circumstances, you may need to enlist the assistance of a qualified electrician to assist you with trickier tasks like wiring and conduit bending. However, you can use them if you possess the fundamental electrical knowledge. Keep in mind to follow all safety precautions!

Let's take a look at some of the typical tools required for installing solar power.

Safety Equipment

Your safety is of the utmost significance because installing solar panels entails heavy lifting, climbing on the roof, and electrical wiring. Consider the following as some of the protective gear you should use.

Boots and gloves

When installing panels, make sure to always use gloves and closed-toe boots to prevent cuts, scrapes, and other harm. Boots with flat bottoms give you the traction you need when dealing with metal, tiles, or roofing shingles.

When working outdoors or on a roof, you can also wear long sleeves to prevent skin burns.

Goggles

When working, put on eye protection glasses and other safety gear to avoid eye injuries. Some sunscreen lotion is also a good idea as you will be working in the sun's direct rays while on the roof.

Rafter anchors

You are prevented from falling when installing a roof-mount system by roof anchors. When working on the roof, always fasten a safety harness to the anchor to give yourself an additional layer of protection.

Scaffolding

You can reach the roof with the aid of a scaffold, which can also be used to lift equipment to lofty locations. You can create a temporary scaffolding system to help with panel installation at higher elevations. Simpler projects can also be completed using standard ladders.

Multimeter

Before you operate on the system, check with a multimeter to see if there is voltage. Before beginning any installation, make sure the wires and terminators have zero voltage; this will prevent electric shock.

Guardrails

You run the risk of falling when installing solar panels on a roof, particularly in larger projects. As a result, makeshift guardrails and other sun safety hacks are necessary. A roof safety strap device also guards against falling.

Ladder

To easily climb up and down, you need a robust ladder. At least three feet of the ladder should be above the margins of your roof.

Power Tools

Portable Drill

A cordless drill is a strong instrument that makes DIY projects easier. It aids in drilling pilot holes, inserting lag screws, and securing them firmly into the roof. Every homeowner should have a cordless drill because they serve so many different purposes. For the majority of solar applications, a general-purpose drill with a 12V or 18V battery is adequate.

Sockets and Drill Bits

Drill bits function as cutting instruments that pierce a variety of materials with holes of varying sizes and shapes. Always select a drill bit that is larger than the hole you intend to make.

You can drill with a manual drill or an electric drill to use the twist bits, which are often used drilling bits for drilling plastic, wood, and metal materials. Masonry bits should be used when drilling into concrete, stone, or brick.

A drill bit socket is a device that modifies sockets for drilling. They have an adaptor on a drill with a tapered shank that is bigger than the size of the produced hole. To tighten the drill bits, sockets are placed on the nuts or bolts as shown below:

Sealant Gun

When screws are fastened to your roofing, this tube of cartridge material (roof sealant) fills up any gaps or holes that may remain. You can utilize a variety of materials, like glass, metal, or ceramic, to link together using the cartridge material, which can be either silicon or latex.

Jigsaw

After installing the solar modules, for example, you can use this effective instrument to cut rails. When cutting intricate patterns or forms, it gives you more control so you can protect your roof.

Saw that reciprocates

With the help of a handheld equipment called a reciprocating saw, you may quickly cut through a variety of materials. The saw has a sizable blade, similar to that of a jigsaw, and an angled grip that make it easy to cut materials on vertical surfaces. On the roof, a powered saw is ideal.

Hole Cutter/Hole Saw

A ring-shaped blade known as a hole saw is used to create holes on surfaces without removing the underlying material. It can be used for drilling, and the pilot drill bit that creates the hole is located in the center to stop the saw teeth from moving. You can cut through thin metal plates, roofing material, and other things with this instrument.

Screwdriver

You can tighten or loosen many types of screws using a screwdriver. Despite the fact that impact drivers have dominated the majority of screwdriving applications, manual drivers may be more practical in some circumstances.

Pliers

Wires can be bent or tiny spaces can be squeezed out if you have long nose pliers. Thick electrical wires can be cut through with pliers that have sharp edges. In order to grip objects while doing the installation, you can also utilize pliers. Using pliers to hold wires helps avoid electrocution.

Tape, string, or chalk line for measuring

It goes without saying that your project need a measuring tape. For instance, to determine where to drill the holes, you can measure the distance between the drilled holes on your panels and mark the relevant locations on your rooftop.

On a flat surface, a chalk line tool draws straight lines. The panels can be installed on the chalk lines by using it to lay straight lines between two spots.

Your solar panels will be put in squares that are absolutely level thanks to a string line. Solar panels on the roof that are organized nicely are appreciated by all.

Tools & Equipment for Wiring

You'll need to understand how to use a variety of tools and machinery designed specifically for wiring.

Boxes and Splices for Junction

An electrical box that houses cables and wires is called a junction box. You need a junction box to prevent wire short circuits if you cut electrical lines or splice the wires together.

By using these safety precautions, you can avoid getting shocked by live wires in your house. A junction box, which will act as an enclosure for all spliced cables, should be used to house all wires related to your switches and lights. A connection splice that can handle two or more circuit cables is part of every box. Cable clamps or conduit connectors are used to secure the cables as they enter the box.

Make sure the junction boxes and splices you buy can handle the voltage and current demands of your circuit before you buy them. Always use insulated lugs or certified splicing hardware.

A/C breakers

When it detects an overload, an AC circuit breaker shuts off the power supply to protect your equipment. When a defect is found, it stops current flow and performs its purpose. Power from your equipment can also be imported and exported by the circuit breakers. A sub-panel with a few circuit breakers is shown below:

Subpanels

When the slots of the main panel circuit breaker are full, this crucial component enables you to install additional circuits in your house. It increases the electricity distribution to particular rooms in your house. If your home or buildings are far from the main panel, it is simple to extend your wiring using numerous branch circuits because a subpanel has its own breakers.

PV metering

You must install a special PV meter that measures the quantity of energy produced by your solar array if you're installing a grid-tied or hybrid solar system. Most of the time, you have to install the meter socket base, and after the utility provider has authorized your solar system, they will install the meter face and turn on your PV system.

Conduits A conduit is a tube used to shield exposed electrical wiring from contaminants, chemicals, and shock. The wires' plastic covering is vulnerable to damage. You can prevent this by pulling individual wire strands into a metal tube to safeguard delicate electrical connections.

A stiff steel raceway known as an electrical metallic tubing (EMT) conduit grounds the panels in addition to protecting the wires. Always double-check your conduit size before beginning the installation process. Once the conduit has been installed properly, a set of channel locks can be used to tighten the joints.

The Channel Locks

Using channel locks is similar to using pliers to grasp, hold, and turn nuts or bolts. Crimping metal objects or the ends of pipes is another excellent use for the device.

Cutters for wire

You can use a wire cutter to cut wires made of iron, steel, brass, copper, and aluminum. The insulated handles offer a secure hold while working and help prevent electrocution from the cables. You can cut cables at an angle that is near to the base if you have a diagonal flush cutter.

Wire Strippers Wire strippers are still another essential piece of equipment since they make it possible to remove the insulation from an electrical wire in order to form a contact.

The wire cutters

A specialized crimping tool known as a hydraulic cable crimping tool is used to crimp wire rope terminals and unite two pieces of metal. Connectors, terminals, and splicers can be crimped together using a hammer-style crimping tool. The crimping tool needs to be pushed with the help of a hammer.

Duct tape

With the help of this practical instrument, electricians may draw a wire through an electrical conduit or route it through a wall. The steel wire used in the fish tape is flat, long, and thin; it is coiled around a wheel. You can link the wire to the fish section and then pull it through the conduit. A conduit is an electrical wiring protection pipe that resembles a plumbing pipe.

Twist Wrench

With the use of a special T-shaped instrument called a torque wrench, you can tighten bolts, nuts, and lag screws on solar panel rails to a predetermined torque. It works perfectly in circumstances when the tightness of screws and bolts is crucial. You can tighten bolts and nuts on the rails with the aid of a high-quality torque wrench. Even though it could take a while, doing this will guarantee that all of your setup's components are firmly fastened in place.

Solar Racking Tools

Roof trusses

A roof rafter is a component of the roof design that extends from the wall plate to the hip (ridge) of your roof. Despite not actually being a solar tool or piece of equipment, it offers base support for mounting solar panels.

The location of the rafters on your roof has a big impact on where and how many panels you decide to put.

Rails

Solar panel rows are supported by rails that are affixed to the roof.

Each solar panel must be installed vertically or in portrait orientation, and two rails with clamps must be used to fasten the panels to the roof or the ground. The rails are fastened to your roof with bolts or screws.

Flashings for Roof Mounts

Roof flashing is a thin metal material that can be made of copper, aluminum, or galvanized steel. Its major purpose is to stop any water from dripping where holes were made in the roof to secure the rails.

End and middle clamps

Mid clamps are positioned in between two panels and keep them fastened to the rails, whereas end clamps secure the panels at the extremities of the rails.

Solmetric Pathfinder/Sun Eye

These methods for solar evaluation or shade analysis provide precise measures of the solar energy produced each day, each month, and each year. These technologies can also quantify the shade patterns present in a certain space.

Tools for Battery Maintenance

Hydrometer

To ascertain the battery's level of charge, this gauges the relative density of electrolytes in a flooded lead-acid battery. A higher level of electrolytes is indicated by a higher sulfuric acid content in the battery. A higher state of charge is produced by a higher density.

Purified Water

Your flooded lead-acid battery's electrolyte level is replenished using distilled water. Depending on the state of your battery, you may need to add more or less distilled water to it. If the battery is brand-new, fill the filler tube with distilled water all the way to the bottom. To the level of the electrode, an old battery needs to be filled with pure water. You can check the electrolyte level in the battery using a little flashlight.

Funnel

This is used to funnel liquids through the lead-acid battery case's tiny entrance so that the battery can be refilled with distilled water.

Rubber Apron You must wear a rubber apron to protect your body and clothing from spillage before handling any poisonous or harmful chemicals. Sulphuric acid or any other potent substance could seriously harm your skin with just a drop.

Nitrile gloves

You must wear gloves to safeguard your hands when handling poisonous substances such as cleaning agents and other dangerous substances.

What is Electricity, and How is it Actually Measured?

Electricity can be defined as the flow of electrons or other charged particles in a circuit. There is energy within us as much as everywhere around us. Since electricity is one of the simplest types of energy to convey across large distances, we use it to carry out work. Humans possess muscle energy, but it is not readily translocated.

Actually, we should start with something a little more fundamental—elements—to understand electricity. Atoms make up every element. The periodic table lists all the elements. Protons, neutrons, and electrons make up atoms. All of us remember the atom from high school. The proton and neutrons that make up its core are surrounded by electrons. All of that, including the electrons and the space between them, as well as the protons and neutrons in the center, make up what is commonly referred to as the atom. Atoms are not static. It's not the same as having a sand grain. The sand grain is a stationary thing, even if it is also composed of atoms. You would notice that an atom is dynamic—constantly moving and changing—if you could see one. When you freeze something to absolute zero, it just comes to a complete stop.

One proton makes up the smallest number of protons in an atom. The hydrogen atom is an atom with one proton in the nucleus and one electron in orbit. It is nature's lightest element. Protons are charged and have mass. Both mass and charge exist in electrons. Neutrons are massless yet have very little charge. Protons have a positive charge, electrons have a negative charge, and neutrons have no charge at all.

The orbiting electrons are drawn to the protons in the nucleus. Another term for the distance between the electron and the proton is the shells that the electrons inhabit. You can visualize the planets that orbit the sun for this part. With the exception of the fact that each electron can fit into more than one shell, each atom's shell resembles a distinct planet's orbit. The valence electrons, also referred to as the outermost shell, are found there. These outermost electrons are the easiest to remove since they are attracted to the atom's nucleus with the least force.

Volts

That electron must move in a specific direction when it is drawn apart and follows a path. An electric field controls that direction. Although the specifics of this are outside the scope of this book, you must remember that the electron will travel from a region of high concentration to a region of low concentration. The potential increases with the size of the gap between higher and lower. The harder the push of electrons in the circuit from high to low, the bigger the potential difference between the two ends. Volts are used to describe that potential. It's common to hear that anything is either 12V, 24V, or 110V. Analogies that use water pressure to represent volts are frequently used to explain this. Consider a tank with two chambers that are divided by a valve. There is around three feet of water in the chamber on one side while there is none on the other. The three-foot column's pressure will force the water into the empty chamber. As the column descends, the pressure will gradually decrease until the water levels in the two chambers are equal. There is no more pushing after that. The two pressures are no longer different from one another.

When a circuit is closed, the electron is almost identically (though not exactly) pushed to travel from the beginning, which destroys the electron in the atom next to it. Then, that electron travels to the nearby atom and displaces the electron in that atom's outermost shell. That keeps occurring all the way around the circuit. The substance is more conductive if an electron may be easily removed from its atom. The less conductive something is, the harder it is to knock an electron out of its shell.

So, give that some thought. What do you use the electricity for when you buy a solar generator? You aren't using the sun's energy to generate power. The electrons in the solar cell are instead knocked off by the photons, and they then go across the circuit. The sun's heat is not used to generate energy. As long as there is enough sunlight, you can produce electricity even in the middle of winter.

When you have a battery, you can determine how much push it has by looking at its voltage. Think of two robots being transported in a trolley. The trolley navigates obstacles and traffic to transfer the robots to various locations throughout the city so they can do mechanical work. The trolley is propelled past impediments and friction before arriving at its destination. The robots dismount, carry out the necessary work, then board the trolley to return to their homes. You can easily push that car if you had a strong bodybuilder, like Arnold Schwarzenegger in his heyday. Pee Wee Herman, on the other hand, wouldn't be able to push too much weight if he were to attempt so. If you have a toddler, it may be a problem if the child even slightly moves the trolley. Voltage is what creates force. It informs you of the maximum load it can carry. The voltage in Arnold would be significantly higher than in the child.

Ampere

Amps are the second topic in electricity that is frequently discussed. Ampere is abbreviated as amps. This is how current is measured. What is hence currently in style? Current is just the rate at which charges are flowing.

If you remember, the movement of charge is what makes electricity. Positive or negative charges make up charges. Electrons have a negative charge while protons have a positive charge. The passage of electrons carrying that negative charge is what we mean when we talk about electricity. Amps count the number of those charged particles passing through the circuit or, more precisely, across the measurement point. That might alter in various regions of the circuit.

A cross-section of a cable reveals that it has a particular area. You can estimate how many charged particles can pass through the conduit from that area. Remember that the movement of charged particles is what causes electricity, and that movement occurs when electrons are knocked off an electron's outer shell. An electron behind another knocks out the one in front of it in a copper wire. The electron is knocked out, and it travels to the atom in front of it. Therefore, if you knew the number of atoms in a specific cross-section, you would also know the number of electrons being bumped at that given moment.

Think about a room full with children. A tunnel that can accommodate exactly four children standing side by side connects one chamber to another identically sized, vacant space. The children would enter the tunnel, travel the required distance, and then enter the second room if they moved along normally. Four children at a time would enter it as they moved along. The room would fill up at a rate of four children per second if each row of children entered once per second. Three rows would enter the room every second if you unleashed a frightful creature to "push" the children through the tunnel more quickly. So, 12 children would enter the room per second. 4 children per second in the first instance, and 12 in the second. The rate of kid movement increases with the amount of back pressure. The monster in the first room is what is making the youngsters go faster—the higher voltage—and the kids are like the charged particles. There is a limit to that, though. If you keep making the monster bigger, you can't expect the kids to run any quicker. There will come a time when the kids' ability to run faster will not be able to keep up with the increase in monster size.

The tunnel's size must now be raised in order to achieve an enhanced flow rate. That would be comparable to expanding the wire's diameter. That would have an impact to some extent. The appliance will be able to draw as much electricity as it needs since it will be possible for more charged particles to flow at once. The more work an electrical device, motor, or other electrically powered device must perform, the more charged particles it will require.

The cross section of the wire has more atoms when the wire's diameter is really increased, which allows it to serve as a larger bridge for the electrons to move over. The number of atoms in a wire increases with its density or diameter, allowing more charged particles to travel down the path.

You also need to take into account the fact that if the wire is lengthy, the distance between the power source and the power usage is greater, and the wire acting as the conduit for the power between them also functions as a resistor.

A resistance is added to the circuit for every meter of wire. Because resistance increases with wire length, more voltage is required to push the charge through. Additionally, keep in mind that if your resistance is great, you will require both more force and charge. Therefore, to push a longer cable, the source's voltage must be increased, and the cable's amperage must be increased to allow for the passage of more charge.

The measurement units for voltage, current, and resistance are volts, amperes, and ohms, respectively. They are in a particular and fixed relationship. The true result of current and resistance is voltage. V is equal to I multiplied by R. In this formula, V stands for voltage in volts, I for current in amps, and R for resistance in ohms. In this relationship, you need to raise the voltage if the resistance is large in order to maintain a constant current flow. You must lower the current if you genuinely want to maintain a constant voltage.

I also equals V divided by R. The only difference is how it is structured in relation to I. Any changes in voltage in this situation must result in an inverse change in resistance in order to maintain I constant. Therefore, a larger resistance demands a higher voltage to maintain the same amps.

Resistance

R is also equal to I / V. Any increase in current in this situation must be accompanied by an increase in voltage in order to maintain constant resistance. You only need to remember this connection. However, there is one part of this that many people would find confusing. Normally, voltage is fixed. Either 110V or 220V comes out of the wall where you live. The voltage is usually fixed because you can't actually modify it. Amps and ohms are variables you can modify. What is pulled rather than what is pushed is more important in amps. It is preferable to picture resistance as immobile, amps pulling, and volts pushing. You can get an understanding of how to control them by using these three states. Change the amount of batteries in a battery-driven system if you require extra voltage. A motor will pull the necessary number of amps if you increase its speed in units of amps. When the voltage stays the same, the additional amp needs less resistance. When you go to increase the amps, however, you see that the wires are beginning to heat up because the wall outlet still has the same voltage. Fuses melt as a result. The filament melts due to the increased heat, causing the circuit to short. You will need to make the cable larger in order to reduce resistance, which will also lower resistance.

The cable's resistance decreases with increasing diameter, allowing more current to flow through it without experiencing a temperature increase.

Always be sure to have a licensed wireman or electrical professional to calculate the length of the wire in the circuit when it comes to homes and mobile platforms like boats, RVs, and the like so that you can end up with the best wire gauge. A fire hazard can result from using wire that is too thin, and that cost will be higher than the initial investment in higher-quality cable. We should also discuss a few other fundamental topics that are covered in my earlier work, which was already stated.

Power

When we use the word "power" in regular speech, it frequently has a different meaning. Power in relation to electricity has a distinct connotation. Voltage and current produce a force that is referred to as power. Remember that volt is the force pushing the electrons and current is the quantity of charged particles flowing per unit of time. Power is the force and amount that arise from particular voltages and amps. 1,320 watts are produced by 12 amps at 110 volts. Voltage times current, or volts times amps, equals power. On a 110-volt system, a 110-watt light bulb is drawing 1 amp. You can get 60 watts from a 5-amp device using a 12-volt power supply. Wattage may be calculated quite easily. Volts—amps—watts.

Amp-Hours

We are aware that the amount of charged particles transferred in a specific circuit is measured in amps. You should imagine batteries as carrying charged particles if you genuinely picture them as bags filled with energy packets. One amp-hour is equal to one amp used for an hour. One amp-hour is equal to consuming half an amp for two hours. One further way to look at this is to consider that one amp equals one coulomb per second. 6.25 x 1018 electrons make up one coulomb. 1 Amp equals 6.25 x 1018 electrons per second as a result. Accordingly, a 1-amp system emits 6.25 x 1018 x 3,600 electrons in a single hour. Accordingly, 2.25 x 1022 electrons move in a single amp-hour. I make this argument in order to make the discussion feel genuine. We feel as though we are being overrun by words and ideas that cannot be put into a picture. However, to help you visualize, think of these as energy packets. Amp-hours are a useful technique to estimate how much energy each of your appliances will consume in a single night and determine whether you can replenish it all the next day.

Electricity Basics and Units

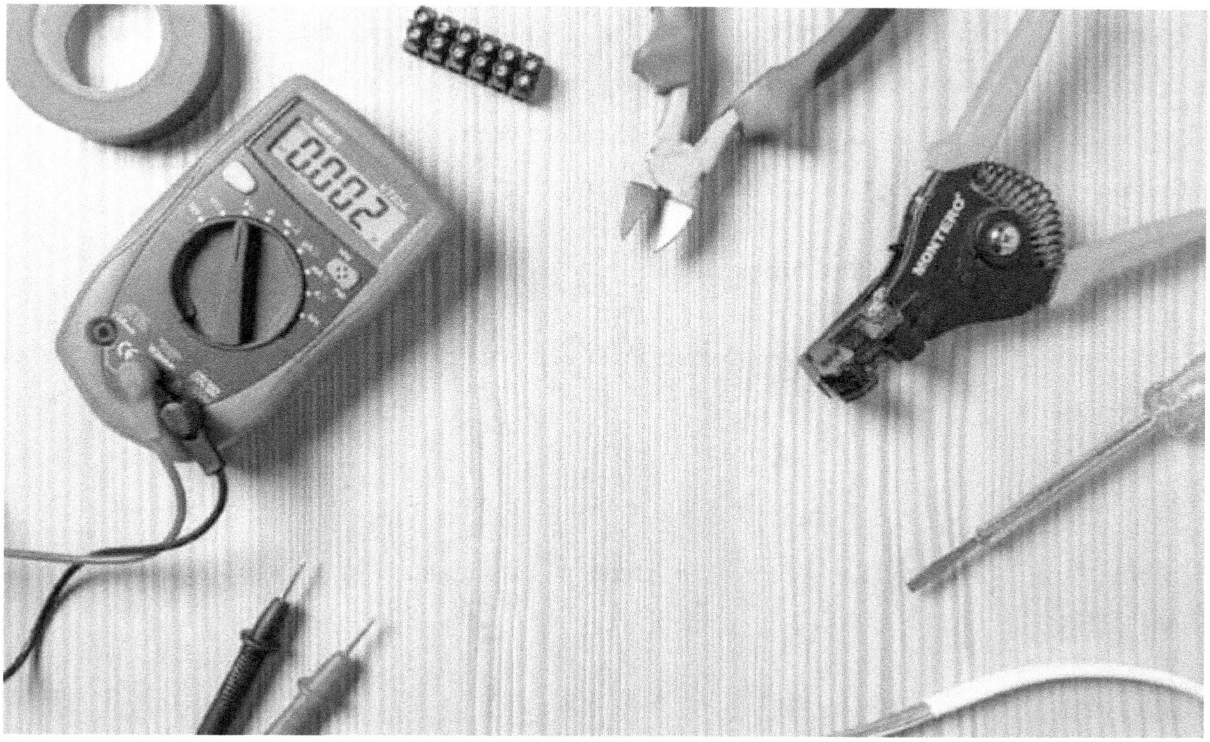

Numerous applications of electricity necessitate an understanding of the underlying principles governing its behavior. You must first understand the fundamentals of electricity, including concepts like voltage, resistance, current, Ohm's law, circuit theory, and other related topics, in order to understand these principles. Clarify what electricity is and is not before moving on. The movement of an electric charge is referred to as electricity. A moving electric charge generates an electric field that can be positive or negative depending on the polarity of the charge. Charge generation is influenced by both primary sources (like solar energy) and secondary ones (like fossil fuels). Whatever the method of charge generation, the age of an electrical current is determined by the passage of this charge. You will learn the fundamentals of electricity in this course so that you may use them more effectively when building and installing your solar PV system.

Circuits and Basic Formulas

Energy-transfer, energy-storage, and energy-conversion components make up an electrical circuit. One or more sources of energy enter the course, and outputs allow energy to exit. Three basic parts make up a circuit: a voltage source, a load, and a resistor. A conductive channel is a conductive substance, whereas a voltage source is a battery that allows current to flow across the course. The circuit's electricity is used by the pack, an electrical device. The load is connected to the power source by the conductive circuit. A fuse is located between the head and the bag and a course has an on/off switch. So educate yourself on circuits and how to put one together before constructing your own solar system.

Simple Electrical Circuit Components

There are several electrical parts in a circuit. You can create your own system with the aid of these components.

Switch

When the circuit is closed, a switch—pushbutton, momentary, or rocker—blocks electrical current.

Resistor

Voltage and current flow in a circuit are controlled by resistors. Before designing the course, the resistor size needs to be specified. Ohm's law can be used to calculate a circuit's resistance. In other words, the voltage and amperage are supplied by the battery and the LED.

Capacitor

Another essential electrical component is a capacitor, which stores electricity and releases it when the voltage drops. It functions as a rechargeable battery would.

Diode

This limits how much power is produced. The main job of the diode is to direct and halt unwanted electrical discharge in the circuit.

LEDs are light-emitting diodes.

The LED is a typical diode that only allows only one direction of electrical flow. When an electrical current runs, the anode (+) and cathode (-) of the LED component produce light. Anode to cathode is always the direction in which electricity flows.

Transistor

When activated by an electrical signal, a transistor functions as a tiny switch that turns the current on or off. In addition, it can boost signals.

Electricity Sources

Any piece of equipment that consumes electricity is a load. Each of these things requires energy and is plugged into an electrical circuit. Electricity is transformed into heat, light, or motion by an electrical load. The term "load" can also refer to the current flowing through a circuit or the amount of electricity consumed by a supply unit. The load is determined by the demand, diversity, power, and utilization of your system. Capacitive, inductive, and resistance.

Defiant Load

Examples of resistive loads include toasters, coffee makers, incandescent lights, ovens, and coffee makers. These devices use active power, which only flows from the voltage source to the load. Additionally, the loads' voltage and current waveforms frequently peak at the same time. In electrical stations, oscillating fixed coils provide sinusoidal waveforms. Resistance and sine wave voltage of a resistive load

Load Inductive

There is no active electricity used by this load. Examples include motors, transformers, and generators. While conducting electricity, the coil of the load stores magnetic energy. This type of load is used by compressors, air conditioners, washing machines, vacuum cleaners, and dishwashers. Some of these devices require peak surge current to operate, even during routine electrical use. Power factor lagging is actually caused by the current waveform, which is actually behind the voltage waveform. In reality, the phases of voltage and current are 90 degrees apart. The voltage is maximum when the current is zero. Reactive power is used by inductive loads, so electricity may flow from the source to the load or vice versa.

Condenser Load

The voltage wave peaks before the current wave in this load. Newer waves always take the lead by 90 degrees. utilized in three-phase induction motors and capacitor banks. because there isn't a capacitive load that is standalone. Capacitors are used in large circuits to regulate power. To improve power factor, capacitors are employed in electrical substations.

How to Select Your Components

Let's quickly review the solar system's setup on a distant platform. Current departs the solar panel as a result of sunlight striking it. It is currently DC current. After there, it passes through a charge controller and finally arrives at the power bank, which is a group of batteries organized in parallel. You need an inverter to convert DC to AC because not all of your equipment will operate on DC. You currently have both AC and DC power at the wall at this time. A schematic showing where the appliances will be installed would be helpful in addition to the list of things you have planned, as many of the equipment you buy will require AC. This will help you determine how much electricity to direct to the installation site.

It's time to consider how to choose your other system components, such as the solar panels, the charge controller, and the inverter, now that you've set aside the system's core. We'll also consider selecting wiring and backup generators.

When buying a big-ticket item, one must strike a balance, as with everything else.

How to Choose Solar Panels

To choose solar panels, you actually need to perform two things. Quantitative in one, qualitative in the other. Quantitative aspects are estimated similarly to how you determined your power bank capacity.
There are three different varieties of solar panels. They all have various cost, usefulness, and longevity profiles. It is up to you to decide what is best for you. In the same way that the cheapest option is not always the most economical, it is not always clear that the most expensive option is the best for you. To determine which options will provide you with the highest return on your investment, you must consider the complete picture.

A solar panel's primary component is silicon, not silicone. The symbol for silicon on the periodic table is "Si". It is the primary component of computer chips and makes up beach sand in its oxide form. It is a common element on the globe with numerous electrical applications, including solar cells.

The three different types of solar panels are amorphous (also known as thin film), polycrystalline, and monocrystalline. Each sort has unique characteristics, making it difficult to categorize them as greatest or worst. Simply put, they are diverse and were created with various budgets and users in mind.

We need to look at this strange aspect of these silicon structures called efficiency. Efficiency takes many different shapes. Efficiency of silicon panels is a little different in this regard. The most effective silicon also happens to require the most expensive purification procedures. The most effective silicon panel reaches a point when its economics render it unviable for more than 90% of the market despite its great effectiveness. As a result, to determine which of the three types of silicon panels is most appropriate for our intended usage, we must look beyond the purifying procedure. Instead, cost and available space are more important factors in efficiency than purity. To put it another way, you should focus on downstream efficiency rather than upstream efficiency. We now turn to mono-, poly-, and amorphous crystalline forms in terms of downstream efficiency.

One can easily tell the other apart from the other. You have probably seen at least one of them, if not all of them. The panels that appear to have had the corners clipped off are monocrystalline ones. If you look at the entire panel, it is made up of several of these that have been arranged in rows and columns. One panel contains individual squares with their corners taken off. Visually, monocrystalline solar panels can be distinguished from polycrystalline ones. The tie structure of the mono panels is absent from the entre panel, which is uniformly organized to resemble a deep blue screen. As a result of their distinct manufacturing processes and placement on substrates that can be either hard (like glass) or flexible and rollable, the amorphous thin film can also be distinguished from the other two.

The most expensive of the three in terms of price is monocrystalline since it has the best efficiency (15% to 22%) and is created from the highest grade of silicon. Since they are more efficient, you need fewer panels overall to produce the same amount of energy as comparable panels made of poly or amorphous silicon. They would yield an output that is about 400% larger than amorphous panels if you were to compare them side by side. Given that solar panels have a 25-year lifespan, the majority of manufacturers offer warranties on them. To gain a fair picture of your expense, you should amortize your charges while estimating them. A better investment would undoubtedly be something that costs twice as much but lasts four times as long. However, it's important to keep in mind that there are two drawbacks. One is the expense, which is obvious—the initial investment can be pricey. The second drawback is the makeup of the electricity generation profile, which is more significant. Warm temperature ranges enable monocrystalline panels to generate at peak efficiency. When it becomes too hot, electrical production starts to suffer. Perhaps monos are not a good choice if you anticipate spending the most of your holidays outside during the summer or if you live in a hot location and plan to use an above-roof air conditioner.

One advantage of thin film, or amorphous solar panels, over the other two types is their ease of production and lower cost. They are much less efficient than mono, which has an efficiency of more than 20%, with a range of 4% to 8%. Despite the fact that they are substantially less expensive than mono, they require five times as much space. The price is not as important as whether or not you would use it in warm weather. A thin sheet is resistant to high temperatures. Your options are also constrained by your footprint, or more specifically, how much room you have on the roof of your van or RV.

Calculate the available square footage before choosing a solar panel, then balance it with the anticipated number of power users. It's a zero-sum game here. Solar panels can be used indefinitely to power all the appliances you require. You will need to use fewer appliances and less energy, or get a larger vehicle to support it.

A Solar Charge Controller to Choose

To stop your batteries from overcharging or overdischarging, you must really utilize a solar charge controller. Occasionally, your solar panels could charge the batteries more than you anticipated (for example, when you travel on a sunny day through snowy terrain), and that extra power from the panels will deteriorate the batteries. You need a solar charge controller to stop that.

You really just need to be aware of two things in order to be able to choose. You must first determine what voltage your system is intended to operate at: 12, 24, or 48 volts? We have been examining 12-volt systems in this book. Therefore, if you followed this guide exactly, the charge controller would be for 12 volts. Some controllers automatically detect the voltage and make adjustments.

You will be interested in the second at peak charging, which is the current. In order to do this, you must consider the solar panels you are buying. In their instructions, they will list peak charging. Add a buffer to the value you see when you search for "short circuit current." If the short circuit current of your solar panel is 8 amps and you add a 25% buffer, you will require a 10-amp solar charge controller.

How to Choose an Inverter

An inverter is the next item you need to purchase. A special one is required for vans, yachts, and RVs. They have ones for house PV systems, but you need to be sure to let the salesperson know if you're installing it in an RV. The second point is that your inverter needs to operate during two distinct periods of use. The first occurs when you turn on a large appliance like an air conditioner or a pump. The peak demand is significant during the first startup phase, but it lasts just a brief period before returning to a stable consumption pattern. You must thus acquire an idea, which you can do using the list of all the appliances and their power consumption that we made earlier in the book. Keep in mind that while the electricity flowing from your batteries is all DC, an inverter is specifically designed for you to be able to run all of your appliances that use AC. The next step is to determine the normal rating after obtaining the peak rating. The load that you will typically exert on it is represented by the normal rating. This is the load that will appear after the air conditioning or the pumps turn on. The average current consumption must be calculated once the transient peaks are reached in order to determine this value.

Choosing Wire

When putting up their system, whether it is their PV system at home or on their boat, or even their hi-fi audio, the majority of people assume that choosing the proper cable is a simple matter. They are unaware that wiring is one of the most popular techniques to increase efficiency and build a better PV system overall. The incorrect wiring can endanger your appliance's safety or worse, decrease the performance of your system.

Having a diagram of your setup and identifying the location of your battery bank is a smart place to start. The size of the solar panels, the length of the cables connecting the solar array to the solar charge controller, and the cabling leading from that to the batteries must all be considered. Cables are also required to connect the batteries in a parallel configuration. Then, connect cables to run from the batteries to the inverter after measuring the cables for DC power to the outlets you anticipate having DC power. Again, wire is required to connect that location to the outlets. Calculating the peak and average loads on each wire is the best technique to ensure that the wiring is done correctly. Therefore, for example, the cabling that connects the inverter to the outlet that powers the 15,000 BTU air conditioner needs to be thicker (lower gauge) so that the appropriate amps can be delivered to the unit. Make the most of the lengths to the outlets you require since wires might be pricey.

Selecting Power Meters and Bank Voltage Monitors

Despite the fact that it is frequently overlooked and misunderstood, a battery monitor is an essential component of the setup. The monitor performs two functions. In addition to letting you know how your batteries are currently performing, it also enables you to schedule future battery charges and discharges. Most of the time, battery monitors are more than just voltage gauges. Voltage monitors are reliable tools for determining the battery's condition. As batteries approach the end of their functional charge, they frequently begin to lose voltage. You can tell when it needs to be charged because the voltage will begin to drop. Many modern monitors can be linked to a computer module, which then immediately launches the software needed to activate your solar panels and start charging your system. Pick one that can connect to a computer or provide a reading to an app on your smartphone. In fact, if you can, you ought to be able to connect each component of the PV system to an app that will let you operate your RV from a distance.

The choice of backup power

It's not absolutely required to have backup power. It depends on your travel habits, where you go, and how frequently you park an RV or a boat at a marina that has electricity. On the other hand, if you genuinely intend to camp out, you'll need a generator for those occasions when the sun isn't shining as brightly or you accidentally consumed more energy than you meant to. A generator is often used for backup power. We'll look at how to choose a generator and what to consider when making that decision.

There are three different types of generators: liquid propane, diesel, and gasoline. Liquid propane would be your first choice, then diesel, and then gasoline, if you are concerned about burning cleanly. However, gasoline is widely accessible everywhere. When you fill up your RV, you can fill it full. The next readily available fuel is diesel, which is followed by the harder to locate propane. The second thing you should consider is whether you are the type who keeps a generator on hand only as a backup for times when there is no sunlight and your use would be seldom. Just keep in mind that whereas gasoline produces more than 125,000 BTU per gallon, propane only produces about 91,500 BTU per gallon. However, gasoline cannot be kept for a long time. They deteriorate, and the fumes can intensify to potentially harmful levels. On the other hand, propane can be kept in storage indefinitely.

Propane would be a better choice if you infrequently use the generator, as we already explained. Gas might be a better option if you frequently use your generators. The fact that not all generators are built the same once you make this decision is the next. One issue is if you only need a portable generator to power a pump. There are, however, generators made especially for vehicles like RVs, boats, and vans.

The generator you require is an RV-class generator. There is no rule, regulation, or law that I am aware of that requires you to purchase an RV generator for your car, but it is a smart move for a number of reasons. Particularly because it has more horsepower and is built with the appropriate ventilation to be stored within the RV.

The startup, or peak power, and the normal, or continuous power, are two ratings that generators also have that you should be aware of. They are known as startup wattage and running wattage in the context of generators. You cannot presume that a generator with a 30,00-watt rating has 3,000 watts of continuous running power; this is startup power. This means that while it can provide a 3,000 watt burst, it cannot sustain that level of power for longer than 10 or 15 minutes. You must consult the instructions to obtain an estimate of the running wattage.

Noise output should be assessed next when considering to purchase a generator. It's important to avoid bothering yourself or your neighbors. After leaving a busy metropolis behind and entering wilderness, the last thing anyone wants to hear is a loud generator churning forth noise and emissions. The noise limit in many campgrounds is 60 dB at 50 feet. Make sure to search for anything that is either better or at least on that level.

Whether the generators are stationary or mobile is another factor to take into account. In most cases, if you have a large RV, you're looking for a permanent fixture that is housed in a space that can be soundproofed and the exhaust can be ventilated all the way to the RV's roof. When it comes to portable items, they are usually lighter and more mobile. When you can store them in the cabin and then use them outside, it is convenient.

The final question is whether or not you can arrange them in parallel, which may be the most crucial of all. You can put the generator between your batteries and your solar panels by connecting them in parallel. You can charge your batteries when there isn't enough sunlight to get them to power up by connecting them in parallel. You can wire them such that you can swap power sources from batteries to a generator if you can arrange them in parallel. This has a distinctive wiring.

First, when the generator and batteries are connected in parallel, your batteries are sent to the inverter, where you continuously receive power. This has the advantage of simplified wiring and is a nice package. However, you won't be able to use the power from your generators to power your inverter without doing some rewiring if the batteries short out or are damaged in any other way. This won't be a problem if you buy quality batteries. The cost of a mechanical switch that sends electricity straight to the circuit breakers and subsequently to the outlets would be higher.
Everything depends on the type of RV you have and the type of RVing you plan to undertake. In either case, having a generator as a backup is wise. You'll be able to stay farther away from civilization for longer periods of time, and it will give you peace of mind when there isn't enough sunlight to recharge your device.

Switch to Disconnect Solar Array

Disconnect switches come in two different varieties. Only one of them is required. The other is for grid-connected fixed installations. The solar array DC disconnect is the first. You require one like this. This is the one that enables you to cut off the power going to the solar charge controller from the solar array. You need this so that the panels won't keep transmitting current to the batteries even when you aren't moving or charging.

Now that you are aware that you require one of these, the next concern is what you should search for while making preparations. First off, connecting a rocker switch between the controller and the array is not as straightforward as it may seem. You must be aware that this is DC electricity, and if you are not using something that is made for it, DC is susceptible to arcing. The panels will send DC through the cable as long as there is sunshine outside, and if a rocker switch is placed, flipping it off to open the circuit could result in a continuous arc. If combined with fumes, that could cause an unforeseen ignition.

Fuses for Your Wiring

Why are fuses necessary for a PV system in an RV or boat? The short answer is that occasionally, there is a potential that the wires will heat up owing to surges in current, an appliance malfunction, or even a short in the wiring caused by mice. However, in the event of a spike, the wires operate as a heated filament and a source of ignition. They could actually be the tiniest part of the total system, but they are unquestionably the most crucial in the event of an accident.

The first thing you should keep in mind is that it's not just about the solar panels' amperage, but also about how you connect them. Although the voltage may be identical if you connect them in parallel, the amps will add up. In other words, if the first panel is supplying 5 amps, by the time you reach the third panel, the wiring is carrying 15 amps. When wiring the panels and choosing the appropriate gauge and fuse, you must keep this in mind. A 12-amp fuse placed on the first section will function; however, if a 12-amp fuse is placed on the second segment, the fuse will blow. One approach to achieve this is to add 12-amp fuses to the solar panel before the combiner, and then connect them using a thicker cable. You won't experience this problem if you arrange them in a series.

Put a fuse after the combiner, after the controller, and after the inverter in the same manner. Fuse each junction using the general guideline of 25% more current than the short circuit current. As a result, the entire system is secure.

Safety Instructions for Handling Solar Power Systems

It is now necessary to acquire safe solar power system operation skills. The easiest installation for your vehicles or house is the solar kit. To prevent casualties, you must take some measures.

Make sure that all of the tools and equipment needed to explore the solar system are on hand. Among them are batteries, charge controllers, mounts, wires, solar panels, inverters, and fuse and fuse holders.

Here is a list of what a solar kit might contain before we learn about safety measures for the solar system.

Solar kits' components

- Solar panel charge controller

- The battery circuit is linked to the solar controller.
- solar panels with a minimum output of 100 or 150 watts
- A washer, a lock, an 11 mm bolt, and an 11 mm hex nut make up the foot mounting hardware.
- black solar panel to controller wire
- Red solar panel to controller wiring
- Tools and equipment for solar installations
- 11-millimeter wrench and a Philips screwdriver
- Sealant that levels itself
- A flush-mounting controller is made with a reciprocating saw.
- 3/8-inch drill and drill bit

Select Your Solar System Installation Method

Make space for all the parts of the solar system to be installed. Thirty feet of wire are included with the package. The wire run from the solar charge controller to the battery is fifteen feet long. The PV panels are connected to the charge controller by an additional 15 feet. Only from the outside is it possible to see the solar charge controller.

However, if you want to mount the panel on a camper, RV, or car, lay down some wood, fiberglass, rubber, or tin. Make sure the panel is facing the sun during the height of the sun's rays if you're connecting it to an off-grid station. Its total efficiency will increase as a result throughout the day. However, their positions need to be changed periodically for best performance.

Map out the path the wire will go by routing it

It is vital to sketch out the direction of the wires from the beginning in order to ensure correct wiring. By doing this, you may also drill the necessary holes and route the cables. The little ring terminals on the wires must be fed from the other ends if the other ends are greater than 3/8 inch. Using self-leveling sealant, patch any holes in the roof. Actually, there shouldn't be any wires running to the battery or solar panel.

The charge controller's holes should be drilled and cut

Any holes needed for the circuit system to work should be cut or drilled. On the solar charge controller, there are symbols like + and -. The battery contains + and - indications as well. These are the links that have both good and bad qualities. Before connecting the negative terminals, connect the positive terminal of the battery to the positive terminal of the solar panel. With a Philips screwdriver, this could be done. Verify that all of the connections on the solar charge controller are tight. The solar charge controller should then be tightened on the outside.

Place the Solar Panel Mounting Feet in place

Using the hardware and an 11-millimeter wrench, attach the mounting feet to the solar panel. The use of two wrenches may be required for this installation. The bolt and lock washer must both be threaded. Pass it through the mounting foot holes and the PV panel holes. The hex nut and flat washer should be connected before tightening the screws.

Using the proper equipment, install the solar panel on whatever surface you like. Keep the mounting feet firmly affixed to the surface by using the self-leveling sealant. Connect the positive wire to the positive axis of the battery using a fuse, and don't forget to cover and seal the tops of the screws at the end of your installations. The actual negative wire should also be connected to the battery's negative axis using a fuse. Verify the tightness and firmness of the wire connections.

Try to connect your batteries in the manner I just suggested if you're using a lot of batteries. Your solar system is now ready to produce electricity after making these connections.

To learn more about setting up the charge controller to regulate the battery and other functions and functionalities, you can also refer to your user handbook.

Guidelines for Stopping and Starting

Always turn off the PV system and make sure the electricity is off before working on any wiring. Residential PV systems of any size can produce more than enough electricity to kill! By turning off the electricity before performing maintenance or repairs on the system, accidents can be averted. Use the correct approach to turn off the full PV system. When working on or near wiring, make sure the power is off with a voltmeter before touching any wires.

While wires are being loaded, never connect or disconnect them. In an electrical circuit, a "load" is anything that draws power from it and causes current to flow. It is risky to interrupt the circuit while the current is running.

For PV systems, disconnect switches and breakers are the main shutdown components. Most systems allow the AC disconnect to be locked out in the "Switched Off" position by locking the switch arm.

Locking out the A.C. disconnect switch stops someone from turning on the power to the house while you're working on the household system. Use the lockout feature on a switch if it has one. It's important to recognize which components of the system are disabled "at zero voltage" and which are not because different systems have somewhat varied shutdown procedures.

For instance, turning off the A.C. disconnect effectively shuts down everything on a microinverter system, with the exception of the solar panels and their wire leads.

By turning on the rapid shutdown disconnect with a string inverter or off-grid system, power is immediately cut off at the disconnecting combiner box that is ten feet from the array. The array is still operational, as is every wire connecting the panels to the combiner box. Covering your solar panel completely so that no light can pass through is the only method to stop it from producing electricity. On the other hand, routine system maintenance hardly ever calls for that.

To stop A.C power input and switch off an auto-start feature, turn off the generator disconnect if the system has one.

Press the shutdown button or turn on the rapid shutdown control box for the P.V. system. Turn off all D.C. disconnects from the PV array if the off-grid system does not include a rapid shutdown feature.

Lock it out after switching the system's A.C disconnect switch to the "OFF" position. This makes sure the inverter isn't working too hard.

Put the inverter's D.C disconnect in the "OFF" position. By doing this, the inverter can't produce AC.

The P.V. charge controller will be turned off if you turn off the D.C. main disconnect.

How to activate an off-grid device

The charge controller or the battery bank breakers will turn on when the D.C. main disconnect is activated.

Put the inverter's D.C disconnect in the "ON" position.

Put the system A.C. disconnect switch in the "ON" position.

Turn the generator disconnect "ON" to start the A.C. power input and activate any auto-start features if the system has one.

Additional Guidelines for Handling Solar Installations

- Work should not be done in inclement weather. Dealing with solar installations in inclement weather, such as during rain, fog, or mist, is not advised. If you must work, make sure you are wearing the proper personal protective equipment, such as non-slip shoes, helmets, and gloves. Once more, the windy weather can make the panels unstable by blowing them off altogether or causing them to move around. The panels can break as a result and fall.
- Do not stand or sit on solar panels. Avoid dropping any heavy items on the panels as well. The glassy covers may break under this pressure, leading to injuries. Additionally, the pressure might harm the panels and result in electric shock.
- Make sure the sheathing on the house is dry. Your roof may leak as a result of this. This issue frequently arises with new construction as well as older homes. If the shingles are taken off during installation, it can occur.
- Make sure the space behind your solar panels is tidy.

- A corrosive environment should not be used to install your solar panels. The International Standard Organization (ISO) assigns the designation C5 to this kind of place.
- Make sure you have lifelines, safety netting, and harnesses if you're working on rooftops.
- Whenever you are working on a solar system, use insulated tools.
- On the solar system, stay away from magnifying lenses and artificial lighting.
- To prevent an explosion and fire, do not put your PV panels close to flammable gas outlets.
- Wearing metallic jewelry while working on solar installations is not advised.
- While repairing the circuit system, cover the panels with opaque materials. This will stop electricity from being generated.
- Do not install your PV panels if you are located less than 0.3 miles from a sea or an ocean. Salty vapors and mists might harm your panels and result in electric shock.
- Install your solar-powered panels on a sturdy roof that can support them safely.
- Make sure the entire apparatus is safely and correctly earthed before mounting a PV panel to avoid electrocution.
- Follow all local regulations regarding the installation of power. You can receive more guidance from your local power company.
- Before utilizing any ladders or scaffolds, make sure they are completely safe.
- Work should not be done in windy or snowy areas. This is to prevent sliding or falling from a height.
- Verify that all of your power tools are operating properly by giving them a thorough inspection.
- Rooftop work necessitates special vigilance.
- Avoid using damp tools, slick surfaces, or working after it has rained.
- Use waterproof fittings or duct seal to prevent moisture from entering the conduit system. The panels will be harmed by this.
- Make sure to install the metal conduit to protect any exposed wires to rain, sunshine, and wind if the circuit system runs through buildings. By doing this, system shocks and short circuits will be avoided.
- Avoid areas where things may fall.

- You should utilize scaffolds if you need to ascend to a height of at least 6 feet above the ground. Ladders should not be used at such heights.
- Partner up with someone who can hold the ladder while you climb up to install the panel.
- From the installed panels, connect a grounding or earthing wire to the surface below. This will shield people from electric shock.

Solar Panel Elements

In order to use solar panels to their full potential, it is crucial to comprehend their variances and the roles that each variety plays.

Solar charge regulators

A charge controller is an electronic gadget whose job it is to regulate a battery's level of charge in order to ensure efficient charging and thereby increase battery life.

The solar charge conditioner, which is installed between the photovoltaic field and the battery, is in charge of managing the energy flow between the two components. The management of the current (I) and voltage (V) parameters during each charging phase is what leads to this control of the energy transfer.

Characteristics of a Charge Regulator

1. The regulator will use the proper algorithm to extend battery life since it is set up for the connected battery.

2. By making up for the greater voltage of the PV field and safeguarding the battery from any overcharges and surges, it also prevents harm regardless of the battery's current state of charge.

3. The regulator may be accompanied by an external display or communication tool, or, if it is a 3-in-1 model, it may be included inside the inverter itself, depending on the manufacturer.

Charge Controller Types

The energy transfer between the battery and solar panel is directed and managed by the charge controller. PWM charge controllers and MPPTs are the 2 types.

Charge Controller with PWM

When the batteries are fully charged, it pulses, cutting off the power transfer between the panels and batteries. It must have the same nominal voltage on the solar panel and battery for it to function properly; for example, if we have a 12V battery, we can only charge it using a 12V panel. Solar panels are created without using all the solar energy and are therefore more cost-effective since they operate with the battery's voltage during the charging phase, an operating point that is not below the maximum intensity that the panel can supply.

In order to avoid raising the voltage, you can connect more 12V panels in parallel if you need more charging power.

A PWM charge regulator loses electricity since it uses the battery charging voltage to operate. The absorption phase, which occurs after the battery reaches the indicated voltage and begins to block the contact between the module and the battery to prevent overcharging, is a significant barrier since it lowers energy efficiency. Its affordability and portability as a result of its modest weight are its strong points.

Charge Control MPPT

They are also known as maximizers since they operate by using the solar panels' maximum output to charge the batteries. This kind of regulator also gets the largest output from the panel, enabling it to run at maximum efficiency, in addition to lowering the current to the battery while it is charging. With excellent conversion efficiency, it internally adjusts the voltage, which is always higher than the battery, to the necessary voltage, gaining strength and reducing total energy output. The increased cost more than makes up for its better production capacity, making it the greatest choice for getting the most out of your solar panel. They can operate with panels and cells at the same nominal voltage, but if we raise the PV field voltage, they operate more effectively.

A DC-DC transformer (which changes HVDC to HVDC as the battery is being charged) and Maximum Power Point Tracking (thus the term) are two features that MPPT regulators have over PWM. In order to maximize power output or restrict it to the sinking or floating phase (phase in which the regulator must maintain battery charge and prevent overcharging and discharging), this regulator works with the module at the voltage that is most suited at any given time.

Due to voltage incompatibility and the presence of more of these outputs, MPPT regulators are more practical because they can accommodate a greater number of PV modules than PWM. However, a PWM charge controller can also be used if we are working with low-power PV.

How Should a Charge Controller Be Selected?

Operating voltage (12, 24, 48) and current (10, 20, 30... amps) are two of the numbers that the PWM regulator delivers. Only at that voltage can we utilize panels to recharge a 12V system. The current and power ratings specified by the regulators that generate our panels in parallel must never be exceeded.

Three numbers are provided by the MPPT regulator: the battery's operating voltage, the panel's maximum operating voltage, and the battery's charge level. The operating voltage at the panel should never be exceeded, allowing us to connect it in a way that ensures that it is always higher than the battery voltage and check the VOC values on the panel's datasheet. The total power of the panel that the regulator can take, however, will be revealed by the battery charge level. At 24V, it will produce 400W if it is 200W at 12V. It may surpass the installed power figures, which would restrict the regulator's ability to charge more power than that.

Last but not least, keep in mind that some panels won't charge the battery properly until you link them in pairs or groups of three MPPT regulators. They may fail to provide enough voltage to properly charge the battery, which causes the battery to suffer early damage.

Follow the manufacturer's instructions and seek guidance if you have any questions, as many solar system failures are brought on by poor connections between the panel and the charge controller.

Sun-powered battery bank

As a backup for cloudy days or nights, the battery pack of the photovoltaic system, also known as the accumulator, is in charge of storing the energy so that it may be delivered at that precise moment without relying on the electrical output of the photovoltaic generator. They primarily comprise of two electrodes submerged in an electrolyte, which when charged and discharged, causes chemical reactions.

Energy Capacity Measurement Unit

The amount of energy that can be drawn from a battery before it is totally discharged (when fully charged) is what truly determines its capacity. For a specific discharge time, its capacity is expressed in amperes per hour (Ah).

The battery's capacity can be specified; for instance, if the capacity is stated as C20, that quantity of energy is the amount that can be drawn from the battery in 20 hours at 20°C until 1.8V of voltage reaches each terminal block of the glass battery. The battery capacity will decrease if it is discharged fast; on the other hand, the battery capacity will grow if the discharge period lengthens and becomes slower. Therefore, a battery with a C100=250Ah capacity can deliver 250A for 100 hours.

Manufacturers typically offer details on the battery capacity at various discharge rates. For computations, the C100 capacity is frequently employed.

Dimensions of Discharge

This is the proportion (given as a percentage) of the battery's charge to its rated capacity. A 250 Ah battery, for instance, would experience a 100% discharge if it underwent a 100 Ah discharge.

The quantity of charges and discharge cycles the battery can withstand at a specific depth of discharge determines how long it will last. Additionally, it relates to the typical depth of discharge. For instance, if the discharge is lowered by 30% in a single cell with a lifetime of 180 cycles and an 80% depth of discharge, the battery's lifetime can be improved to 1000 cycles.

Expected lifespan

In most cases, life expectancy is expressed in years. However, the amount of cycles a battery has determines how long it should last, according to the manufacturers. Batteries lose capacity over time, and even if they are still functional, when they have lost 20% of their initial capacity, they are said to have reached the end of their useful life. It must be considered from the very beginning of the system's design.

Dimensions of discharge The DOD (depth of discharge) measurement is the capacity utilization rate. Battery longevity (number of charges and discharges) versus DOD (% of the battery's capacity). A shallow cycle battery with less than 25% DOD is predicted to last at least 4000 cycles. The battery's lifetime would be lowered to 1500 cycles if it were utilized but cycled at 80% DOD (1 cycle = 1 day).

PV Inverters

The direct current produced by solar panels is converted into alternating current by solar inverters, sometimes referred to as photovoltaic inverters. This alternating current can then be used in homes, stored in solar cells, or fed into the grid.

However, when we discuss these systems, we typically just consider the solar panels that are mounted on roofs or other surfaces. Solar systems also require other components to ensure their functionality, with inverters standing out.

This device, often referred tc as a photovoltaic inverter, can transform the power generated by the solar panels for use in the house, studio, or safe haven.

The solar panel's DC output must be converted to AC, which is needed by all commercial equipment, thus the name "inverter."

How Solar Inverters Operate

Solar panels are in charge of capturing solar energy and using the photovoltaic process to turn it into electricity. Crystalline silicon or gallium arsenide semiconductors are used to make the positive and negative layers of solar panels, which are joined by junctions. These layers collect solar energy and transfer it to the photovoltaic cells.

But since this electricity is variable and direct current (DC), a solar inverter is needed to change it to alternating current (AC). In order for the numerous appliances and electrical devices to operate as efficiently as possible, the inverter must supply the 230 V standard that is used in Spain.

Solar inverters perform a number of crucial tasks in addition to actually turning direct power into alternating current, including:

The inverter's job is to use all the electrical energy produced by the solar panel while also stabilizing the electrical wave to ensure a stable power supply within the designated range, protecting household appliances and other electrical and electronic equipment from damage. This will improve the installation's performance.

Protection and command: The inverter continuously tracks system performance and evaluates energy efficiency. It is in charge of detecting any event and alerting the user to potential mistakes. It also dissipates the heat produced.

What categories go under solar inverters?

Grid-connected or off-grid photovoltaic systems are categorized differently in solar inverters.

Inverter Connected to the Grid

Solar inverters must be synchronized with the conventional grid in order to connect solar installations and produce or sell energy. Grid-connected inverters come in three varieties, and they are available on the market right now:

Inverter attached to a string The most typical kind of inverter seen in self-use systems is this one. The electricity produced by the solar panels is sent straight to the inverter because they are connected in series. The panels are quite effective when there is no shade present, and optimizers can be placed to further improve efficiency.

Inverter from strings to grids with optimizer The optimizer is affixed to each solar panel or built into it. The efficiency of the system is increased by sending the energy generated by each panel to the string inverter. Additionally, each solar panel's performance can be checked on its own.

Microinverters or distributed inverters: As the PV system will still produce power, even though certain solar panels may have a lower yield owing to shading, they are installed separately on each solar panel to minimize the detrimental impacts of shading. Additionally, they make it possible to track each panel's effectiveness separately.

The Off-Grid PV System Inverter

Batteries are required since this inverter is utilized in photovoltaic projects and is not connected to the public grid. They are mostly utilized in distant or maritime photovoltaic power producing systems.

They are in charge of converting battery current to 220V so that appliances and other electrical or electronic devices can be powered effectively. When the voltage drops too low, they are programmed to switch off the power. Additionally, they permit the installation of a charger that kicks in when the voltage is too low to start charging the battery.

Blended Inverter

They are made to gather power from solar cells and the public grid. They are utilized in locations with challenging power supplies.

Cables and Connectors

The MC4 enables modules to be joined, making it simple to form panels (or strings) by manually pushing the connections of nearby panels together. To avoid being unintentionally detached by pulling on the cable, they must be unplugged using a specialized instrument called an MC4 connector wrench. Since 2011, MC4 has been a standard feature on all solar panels manufactured.

Design and Organization

In reality, the MC4 system is a plug-and-socket setup. A plastic enclosure encloses the male and female connectors. The right cable diameter is required by the MC4 for good sealing. Double insulation and UV protection are usually required because most cables will deteriorate if used outside without sunshine.

A pair of MC4 typically comprises of 10 components in total, 5 pieces per connector, as seen below:

Important details

It is crucial to avoid connecting or disconnecting while under load, even in low voltage systems (12–48 V). Due in part to direct current (DC), arcing can occur and cause serious damage to the materials in contact, leading to high resistance and overheating. In contrast, if alternating current (AC) is used, the arc will self-extinguish at the point where the voltage is zero.

Not eligible are plug connections that are not produced by Multi-Contact (MC), but which are compatible with the manufacturer's Multi-Contact (MC) products and are occasionally referred to as "MC Compatible" by their maker. It is not advisable to link long-term safe and steady electrical connections for security concerns. In the event that these unapproved connectors are coupled with MC components, MC disclaims any liability.

A breaker should be installed close to the connection to open the circuit without the risk of arcing when the MC4 is detached. MC4 connections are made for conductor diameters of 12 and 10 AWG (4 and 6 mm2) and have a 30 A current rating.

There are two fundamental tools needed to install the MC4 connector correctly. The MC4 punch down clamp, which is used to clamp the pins, is the first necessary piece of equipment. It is crucial because it applies the right amount of pressure to attach the pins to the cable.

Because of the possibility of arcing brought on by loose pins, using standard jigs or just manually bending the retaining pins is dangerous.

The second tool is an MC4 wrench, which is also used to disconnect the MC4 terminals. It tightens the terminal cap to secure the connector.

Solar Battery Bank

Off-grid solar power systems, usually referred to as "standalone" systems, are created to run separately from the grid. Since your solar panel does not produce electricity at night, the system needs to store some extra energy in the batteries in order to maintain power flow.

In PV systems, solar batteries perform their function by storing the energy produced by the solar array. While some newer battery technologies have integrated inverters that offer integrated energy conversion, the bulk of current systems send electricity from the batteries to the inverter or inverters so that it can be converted into AC power that is suitable for AC equipment.

When your array's energy output exceeds your consumption requirements during daylight hours with strong sun exposure (about 10 AM to 2 PM), batteries can aid by storing the extra energy. More power can be stored the larger the battery's capacity.

A Deep-Cycle Battery: What Is It?

Like all solar choices, deep-cycle batteries are made to offer consistent production for a long time before being discharged to their recommended limit. This kind of battery is distinct from automotive batteries, which must constantly be prepared to produce a burst of high power output, for example, to start the car engine. Off-grid and hybrid PV systems can only use deep cycle ones, as this definition suggests. Solar systems can be used with a variety of deep cycle battery technologies, including lead-acid, lithium-ion, nickel-based, and flow batteries.

Various battery types

The two most prevalent battery chemistries for off-grid systems are lithium-ion and lead-acid. They do differ in many ways, including cost, longevity, and capacity, in addition to their chemistry.

With the exception of cost, lithium-ion technology is more recent and superior to lead-acid in practically every respect. Despite having better efficiency, capacity, depth of discharge, and lifetime than their more traditional rivals, they are, sadly, much more expensive.

Other varieties of solar batteries include nickel-cadmium and flow batteries. They won't be discussed here because they are seldom ever used for residential or recreational reasons.

Deep-cycle lead-acid batteries

For over a century, energy has been stored using this technology. These batteries are still incredibly well-suited options for small and medium-sized PV systems because of their dependability and inexpensive price.

The biggest drawback of these batteries is their upkeep, particularly with flooded kinds, which necessitate periodic refilling and ventilation. Installation problems will also result from ventilation considerations. Other drawbacks include the depth of discharge and their shorter lifespan (seven to ten years). To avoid irreversible damage, these batteries should only be drained up to 50% of their capacity. Your best possibilities for a lead-acid battery might be Deka or Crown Battery.

Lead-Acid Battery Mechanics

One or more of the following compartments make up a lead-acid battery:

1. Lead sheets

2. Electrolyte, a diluted sulfuric acid solution

3. Anode, the negative electrode

4. Cathode, the positive electrode

The lead plates serve as a divider, preventing a direct connection between two diametrically opposed electrodes. Lead oxide is formed at the cathode during charging, whereas water is produced during discharging as sulfate ions leave the electrolyte. The combustible gas hydrogen may develop if the battery is overcharged. Due to the increased risk of an explosion, it is crucial to properly monitor and ventilate the batteries.

Lead-Acid Battery Types

- Traditional deep cycle solar power systems, golf carts, and motorbike engines all start with flooded batteries. The user can open the lid and add distilled water as the battery runs out.

- The same principles apply to sealed batteries as to flooded ones, but before the battery is sealed, a sufficient volume of liquid is supplied. You cannot add distilled water to this type, as the name suggests.

- Lead-acid sealed batteries having a regulating device to allow oxygen and hydrogen gas to safely leave the cell are known as valve regulated lead-acid (VRLA) batteries.

- Absorbed gel matt (AGM) lead-acid batteries are a variation of VRLA batteries built using a more sophisticated technique for maintaining electrolytes suspended around active plates to enable a respectable depth of charge and discharge. Among the uses of this kind are solar and storage.

- Gel lead-acid batteries use the same electrolyte suspension method, despite the fact that they are not wet cells because silica has been added to the components. While they have slightly longer lives in hotter climates, there is a greater likelihood that they will have shorter lives if charged at the incorrect voltage.

Features and Terminologies for Lead-Acid Batteries

- Let's first go into basic terms to get familiar with the subject. For optimal use and maintenance of your PV system battery bank, you might need to employ some of them.

- The total capacity of a battery is its capacity to store the most energy, expressed in amp hours. Half the capacity of a 200-amp hour battery is offered by a 100-amp hour battery.

- The ratio of the density of the electrolyte to the density of pure water is how specific gravity is truly determined.

- The resistance brought on by the battery itself is referred to as internal resistance.

- The current produced by the battery at 32°F for 30 seconds while the voltage is 7.2 volts is known as a "cranking amp" (CA).

- The term "cold cranking amps" (CCA) refers to the amount of current produced by a battery at 0°F for 30 seconds while maintaining 7.2 volts of voltage.
- The amount of time in minutes that a battery can discharge 25 amps at 80°F while still staying over 10.5 volts is known as reserve capacity (RC).

The amount of battery storage that can be safely used up before the battery is damaged is referred to as the depth of discharge. In truth, the recommended depth of discharge for lead-acid batteries is at about 40%. Lead-acid batteries will suffer severe damage if discharged more than 50%. This will ensure a long lifespan, but to meet the complete electrical requirements of your equipment, you will need a larger battery store.

The pace at which a battery discharges its whole capacity at 25°C over a specific amount of time (five, ten, or most frequently, twenty hours) is known as the C rate (load factor). For example, the discharge current of a battery with a 20-hour rating of 100 amp-hours will be as follows:

100 amps per hour divided by 20 hours equals 5 amps as the discharge current.

The battery power is depleted after 20 hours at this current (5 amps), which is the discharge current. Now, we can calculate the C rate by dividing the current of discharge by the total amp-hour of the battery:

C rate (load factor) equals 5 amps per 100 amp-hours, or 0.05°C rate.

The most common batteries on the market have nominal voltages of 12, 24, or 48. The storage capacity of the system is not based on the voltage of a battery; rather, it depends on the voltage of other PV components.

Sulfation of Lead-Acid Batteries

Sulfation, a process in which sulfur crystals form at the lead plates and stop the necessary chemical reaction from occurring, is the most common reason for battery degradation and short lifespan. The battery's low charge or electrolyte level is typically to blame, therefore you should always keep a watch on these two crucial characteristics.

Keeping an Eye on a Flooded Lead-Acid Battery

It's especially important to keep an eye on the two aforementioned factors in flooded lead-acid batteries. A flooded lead-acid battery must be monitored and refilled with the following instruments:

1. purified water.
2. a voltmeter to determine the battery's voltage.
3. a hydrometer with temperature compensation can be used to calculate the battery's specific gravity.

4. appropriate safety goggles and gloves.

The electrolyte level and charge level of flooded lead-acid batteries must be regularly maintained through refilling; in contrast, with AGM lead-acid batteries, these two elements cannot be changed, making AGM lead-acid batteries less prone to harm.

Fluid level and state of charge are two elements that significantly affect a battery's health, as was previously mentioned.

Liquid amount

You can examine the level of the mineralized (distilled) water by removing the flooded lead-acid battery's cover. Even though most batteries have a fill level, if no metal lead surface is visible, you can make sure you have adequate water. Never overfill the battery; the recommended level is often half an inch below the cap. Always keep the aforementioned safety precautions in mind.

Charge state and discharge level

We can quickly determine a lead-acid battery's depth of drain and state of charge by measuring its voltage and specific gravity.

The data for a 12-volt battery's depth of discharge in relation to voltage and specific gravity are shown in the table below. The numbers are readily multiplied by two to account for a 24-volt battery.

Considerations for Charging

Before assembling your battery, it's crucial to understand the three primary charging phases:

1. Float or trickle charging keeps the battery completely charged by charging it at the same pace as it is being discharged.

2. When recharging a fully drained battery, bulk charging is required. To obtain the highest voltage advised, the voltage is raised.

3. Voltage is maintained at its maximum throughout the absorption phase, which comes after the bulk phase, while amperage decreases until the battery is fully charged.

The best results are obtained with adjustable charge controllers adjusted for the aforementioned phases and in accordance with the instructions provided by the battery's producer. Bulk and absorption charging times are often quite brief in small off-grid systems, making it simple to set the same voltage for both phases.

Effect of Temperature

In order to charge a lead-acid battery in temperatures that are higher or lower than ideal, additional considerations must be made, the most crucial of which is the voltage adjustment. In colder weather, these batteries need to be charged with a greater voltage, and in hotter weather, with a lower voltage.

Lead-acid batteries also lose capacity as they become colder, and load and charge rate are also impacted. Your storage capacity may be significantly impacted by these elements. larger temperatures would also increase the possibility of larger currents, which would present a fire hazard; accordingly, certain factors must be taken into account when sizing your battery bank and system.

The Peukert Law

Lead-acid batteries' capacity decreases in colder climates. Peukert's law explains why this is. Peukert's law has developed an exponent that shows a battery's actual discharge time based on internal resistance and recovery rate. The real discharge time increases with decreasing number. The Peukert number of lead-acid batteries ranges from 1.3 to 1.5. Let's examine its operation:

In order to determine the actual discharge time, you must also divide the total discharge time by Peukert's exponent, in this case, 1.3 (5 / 1.3 = 3.8 hours, the actual discharge time), for a 100-amp-hour battery with a 20-amp discharging rate. Temperature can also have an impact on these numbers. The image below demonstrates how increasing temperature both increases capacity and decreases C rate (also known as discharge or load factor).

Temperature can endanger the battery, shorten its longevity, and increase the risk of fire. Lower voltage is needed for the given charge current as the temperature rises. In order to make up for the decreased voltage and maintain the equation's constant value, the current increases:

$V = I * R$

If not managed, the increased current and heat from the charging process may result in a state known as a runaway condition, which can start fires.

varied lead-acid batteries demand varied changes due to various production processes, particularly with relation to charging voltage. In the specification sheet for lead-acid batteries, the amount of temperature compensation voltage is readily available. Your charge controller needs to be set up in accordance with the temperature compensation voltage and the environment where the battery is installed. Remember that each cell has two volts, therefore a 12-volt battery is made up of six cells (12 volts/2 volts).

Alternately, a drop by roughly 3 millivolts per Celsius degree below 25°C is advised if a consistent charging current is wanted.

VRLA and FLA, two popular types of lead-acid batteries, require various charging and maintenance requirements. The primary variations between these two battery types' maintenance requirements are shown below.

Battery Charging and Maintenance: VRLA versus FLA

Of all lead-acid battery types, flooded, unsealed batteries require the most upkeep. They must be replenished periodically. They require a properly ventilated area, therefore connecting the battery box to an outdoor area is a smart option in terms of ventilation. Ventilating fans must be fitted in larger installations. As previously stated, this is because these batteries, particularly when overcharged, have a tendency to release hydrogen.

However, because the manufacturer has sealed the VRLA batteries, they can never be recharged. Although less severely than for FLA batteries, ventilation care is still necessary because they only release gas if overcharged.

A hydrometer should be used to measure the specific gravity of FLA batteries to assess their level of discharge; however, due to the sealed nature of VRLA batteries (as well as AGM and GEL), this is not possible. Among lead-acid batteries, FLA batteries have the broadest tolerance for voltage. As a result, they need to be connected to a suitable charge controller with a suitable charging schedule.

Compared to FLA batteries, VRLA batteries are significantly more voltage-sensitive. They require adequate charging if they are to have a pretty lengthy lifespan. The battery becomes hot and releases gas at higher voltages than are safe, permanently damaging the battery. In contrast to FLA, where it is significantly less dangerous, this scenario could result in the battery suffering permanent damage from hydrogen emission. The secret to having a properly maintained VRLA battery is choosing the right charge controller.

Lead-acid flooded batteries (both sealed and unsealed) cannot be positioned on their side due to the possibility of leaking, as suggested by their name. VRLA batteries can be installed with more flexibility because they can be oriented on their sides.

Battery, lithium-ion

Lithium-ion technology has been discovered to be highly beneficial and dependable in relation to storing electrical energy as electric vehicles become more and more common. These batteries are now often utilized in virtually every cordless electrical equipment, including laptops, cell phones, and other devices.

The first manufacturers of the lithium-ion batteries used in residential solar storage systems were Tesla Powerwall and LG Chem.

While lithium-ion batteries work best for larger systems, lead-acid batteries are better suited for small, little used off-grid devices. This newcomer is the best alternative for residential uses because of its longer lifespan, deeper discharge, superior efficiency, and larger capacity.

Lead-acid batteries cost far less than lithium-ion batteries. The only drawbacks of lithium-ion batteries are their higher cost and heat-releasing chemistry. The thermal runaway feature of these batteries, which releases heat, can significantly increase the likelihood of a fire starting, especially if fitted incorrectly.

Mechanics of Lithium-Ion Batteries

The following parts are found in every lithium-ion battery:

- Cathode
- Anode
- Electrolyte
- Separator
- Two collectors of both negative and positive current

While lithium ions are stored in the anode and cathode, they are transported from the anode to the cathode and vice versa through the separator and electrolyte. As a result of this movement, the anode will release unbound electrons, creating a charge at the positive current collector.

The battery-powered device will conduct the electrical current from the current collector to the negative collector. It is the job of the separator to stop the flow of electrons inside the battery.

Lithium-Ion and Lead-Acid Battery Comparison

Cost: While a 12-volt lithium-ion battery with a 100-amp-hour capacity costs about $1000 USD, a comparable VMAX lead-acid battery costs about 1/4 of that amount. Given that lead-acid batteries have a 50% discharge limit as opposed to 100% for lithium-ion batteries, you would need two lead-acid batteries to store the same amount of energy as one lithium-ion battery, but you would still pay half as much for them.

Efficiency: It has been established that lithium-ion batteries are more efficient than lead-acid batteries. Lead-acid batteries have an efficiency range between 80 and 85%, but lithium-ion batteries are thought to have a 95% total efficiency. Greater usability of stored energy can be achieved through higher efficiency.

Capacity and depth of discharge: Lithium-ion batteries store more energy than their more traditional competitor with the same amount of space thanks to contemporary technology and their chemistry. Overall, less space is required as a result. This will be a good feature, particularly in bigger off-grid setups where batteries take up a lot of room.

The proportion of battery storage that can be safely used before the battery is damaged is referred to as the depth of discharge, as was previously explained. Unlike lithium-ion batteries, which have 85% or even more of their capacity available for use, lead-acid batteries can suffer catastrophic damage if they are discharged by more than 50% of their overall capacity. The word "effective capacity" is used to define the depth of discharge in conjunction with the battery's capacity.

Battery lifespan is determined by how many cycles of charging and discharging they can withstand before suffering major harm. Compared to lead-acid batteries, lithium-ion batteries have a far longer effective lifespan.

Choosing the Type of Battery

Considering all we've discussed, it's clear that lithium-ion batteries are priceless due to all the qualities we've listed; nonetheless, lead-acid batteries are the best option for small systems and for people just getting started with solar energy who wish to dip a toe in the water. Lead-acid batteries are the ideal option for people on a tight budget because of their relatively low upfront cost.

Cost-effective lead-acid batteries can also be useful for cabins, yachts, and RVs that aren't used very often. All the aforementioned benefits of lithium-ion battery technology are applicable to residential battery-backed systems, especially if a more economical and space-efficient PV system is desired.

Solar power systems are gradually becoming inseparable from batteries. Knowing the various battery kinds and functions can help you get the most out of your PV system because new battery technologies have made them more affordable than ever.

Installation Guidelines

It is advised that the homeowner buy all of the aforementioned parts rather than making them. Make sure to build a risk-free system and abide by local legislation. Determine whether a PV system installation was successful or unsuccessful by following these guidelines:

Personnel

- Full installs are more appropriate for team efforts because of their complexity.
- You can get in touch with a solar provider, an electrical subcontractor, or an electrician.
- Compared to other people, company professionals have more experience with PV systems.
- The group must plan its services and implement them in accordance with building codes.

- Inspections, meetings, and warranty assurance are necessary for solar power installations.

Norms for Safety

- Anything with high voltages and electricity needs to be safe.
- At all costs, the installation crew must safeguard the property, occupants, and machinery.
- Electrical systems include the circuits and solar panels.
- The PV panel configuration must adhere to the electrical safety regulations in your application location.
- It is important to safeguard everyone's health and safety during the installation.
- Remove the risks and electrical hazards related to climbing tall roofs.
- For their own safety, the installation specialists must to adhere to the best practices.

Power Rules

- Fixtures for direct current (DC) wire panels are frequently rather big in size.
- Outdoor wiring should be shielded from extreme temperatures, wind, and rain by the business doing the installation.
- For conduits, cabling, and electrical wiring, one can drill the roofs.
- Utilize the assistance of building professionals to ensure that the roofs are sealed off.

- Depending on their requirements, System Design One can install a pre-engineered panel system.

- Size each component with care to create the ideal PV system.

- Considerations such as the building's gable, roof pitch, and shade must be made.

- Design the grid-connected and off-grid systems precisely.

- Distributors, traders, and solar companies can also plan and carry out.

Costs

- Software tools are available for cost estimation and design.

- In terms of energy, one can figure out the installation costs "per watt".

- In PV panels, cost and efficiency are inversely correlated.
- The complexity of the task (such as the type of roof and building architecture) will determine the labor costs.

- The price goes up if the solar power system makes use of all the parts.

How Many Panels Do You Need?

The most cost-effective and environmentally friendly form of energy currently available is solar electricity. You may live off-grid and still have access to all your electricity needs if your home is equipped with the correct solar power system. If your solar power system produces more energy than it needs, you can feed it into the grid and make some money. However, there is a problem when figuring out how many solar panels you would need for your home.

How to Determine Usage

Prior to figuring out how many solar panels you will actually need for your home, you must first understand your objectives. An excellent strategy to minimize your carbon footprint and increase return on investment is to invest in a specific solar power system. The initial cost of installing a solar power system at your home may be high when you purchase the necessary hardware, but you will receive free energy from the sun for the duration of the system's useful life.

You must know how much energy your household consumes in order to determine how many solar panels you actually need, as you will see in more detail below. You must also take into account the area of the roof where you will put the solar panels because they come in various sizes. You need to determine the peak sunlight and environment where you reside because solar panels offer various degrees of watts. The quantity of photovoltaic (PV) panels you'll need for your house depends on the efficiency of the panels you choose.

If you want to acquire the correct information regarding how many solar panels you will actually need, consulting a competent solar installation is crucial. A specialist will assess your household's energy requirements and determine how many solar panels you would require.

What Kind of Solar Power Is Required?

They will assist you in figuring out your home's typical energy needs. For instance, divide your home's hourly energy demands by the solar panel's wattage and increase the number of solar panels you need by the highest number of sunlight hours in your area. The amount of solar panels you may need on your roof will depend on the wattage each solar panel produces.

The effectiveness and quantity of solar panels you will need to power your home depends on how much sunshine hits your roof. Different specs and features of solar panels have an impact on their effectiveness. The precise number of solar panels you should install on the roof will depend on a number of additional considerations, including battery storage and roof size.

How to Calculate the Watts You Need

To find the 'Kilowatt Hours' (kWh) utilized, look at your average electricity bills from the prior months. First, note the duration, which is often displayed on your electricity statement as 30 days. To determine the number of kilowatt-hours consumed over a certain time period, you can also utilize the method of subtracting the prior meter reading from the present one.

For accurate figures, you'll need hourly and daily consumption. You can either manually calculate the kWh for your family use or acquire this information from your power bill. You can consume roughly 200kWh every month, depending on the area and size of your home. Larger homes, on the other hand, need 2,000 kWh or more of energy for air conditioning. However, the average US home consumes 900kWh per month. This corresponds to roughly 1.25 kWh each hour or 30kWh per day.

Therefore, to determine the daily average solar energy requirements for your home, consider your average daily energy consumption as your aim. Your overall energy requirements should be used to calculate the amount of kilowatt-hours your solar power system will need to generate. It is crucial to remember that not all solar panels perform at their highest level of efficiency all of the time. Your solar power system's effectiveness may be affected by factors like the weather.

On overcast or wet days, your solar power system does not provide the most electricity. Other strong panels can still work in these conditions, but they won't be as effective as they are when it's sunny. In order to calculate the total number of solar panels you could require, you must look at the average wattage provided by each panel. To receive the correct information about the solar panels that you can purchase for your home, speak with a qualified solar panel installation.

How much sunlight is there in your area?

The amount of sunlight you anticipate receiving in your region is a crucial factor as well. Your solar power system's capacity to produce energy on a given day may be directly impacted by your peak sunlight hours. Compared to other destinations, you can anticipate more peak sunlight hours in some localities. You can still use solar power even if your location has less peak sunlight hours; however, you will need more solar panels to do it.

The Renewable Resources Data Center has accurate information about the amount of sunlight in your area or state, which can be used to assist you make decisions. You should multiply your solar power's hourly generation by 1,000 to get watts. Additionally, you must divide the average daily peak sunlight hours in your location by the average hourly wattage requirements. You'll learn how many solar panels you could need to generate energy hourly from this. You will need 6,000 watts in an area where a residence uses 900kWh on average per month and has five prime hours of sunlight each day. This indicates that you will require 20 300 watt solar panels. You will require 40 140-watt solar panels for your installation. To get the right amount of panels, you must consider the size of your roof, though.

Other Elements that Impact Solar Energy Efficiency

The efficiency of solar output varies depending on a number of variables. There are variations among solar panels. For instance, the typical photovoltaic (PV) solar panels found on residential buildings range in wattage from 150 to 270 watts per panel. Efficiency refers to the rate at which each solar panel can transform solar energy into light, and is determined by the size of the panel.

The efficiency of various solar panels is additionally influenced by cell technology. Gridless solar cells can absorb more sunlight than other types of conventional cells. Due to their strength and resistance to corrosion and cracking, panels without grid lines also do not face problems like delamination or peeling. In addition, compared to a single inverter located within the home, a microinverter mounted on each panel aids in optimizing the conversion of power at the source.

Since they vary in effectiveness and quality, it might be challenging to choose the best panels for your home because of these variables. The differences may change how many solar panels you require for your home. You will need solar panels on your roof since more effective solar panels can generate more electricity. Less solar panels that are more powerful and efficient can nevertheless provide the same amount of energy. To determine the ideal amount of solar panels for your home, look at the wattage that each panel can provide.

In order to get the precise amount of solar panels you need for your property, simply divide the hourly wattage needs of the building by the wattage of the panels. About 20 solar panels with a 300W output are required for a typical-sized residence. But you will require more solar panels with lesser wattages.

What impact does the size of a solar panel have on energy production?

The size of each solar panel you select for your roof will be a major factor in how many panels you ultimately require. You can purchase larger solar panels for less money if your roof is bigger, but doing so may reduce their effectiveness. To maximize the use of solar energy on a tiny roof, it is a good idea to utilize fewer solar panels with high efficiency.

The most typical size of a household solar panel is 65 inches by 39 inches. However, the sizes of solar panels vary depending on the manufacturer, so you should do your research before deciding. The efficiency and energy production of other panels have altered throughout time, even though their size haven't changed considerably. You can reduce the overall number of solar panels you would need on your roof by choosing ones that are more efficient than others. You can determine the precise quantity of solar panels you need for your home by considering your electricity generation demands.

A qualified solar panel installer must first examine your roof to evaluate its structure and the sun's angle. Your home's location will also influence how you set up the roof's solar panels so that you can meet your daily energy needs. Another aspect to take into account while installing solar panels is net metering. This is in reference to how your utility company will credit you for any excess solar energy that your system generates during the day. If there is not enough light to create power, especially at night, your utility company will allow you to draw from the credits when you switch to a traditional power grid. However, how well your home's solar battery system performs will determine this. Make sure your appliances won't overload your home's energy supply system.

Considerations Regarding Water Pump Load

There are various aspects to take into account when determining how much solar energy you will need for your water pumping equipment. Numerous elements that affect the solar water pumping system might also determine how many solar panels you require. For instance, figure out how much water each person in your home needs and multiply that number by the number of occupants. This will enable you to select the best solar pumping system for your household's requirements.

The amount of water the pump can move per unit of time, or its pumping rate, must be taken into account. You should also take into account additional friction losses that can occur along the pipeline as well as the height at which the pump must operate in order to convey the water. The number of hours the pump can run per day must be taken into account when calculating the amount of fluid a pump powered by solar panels can move per minute. You should also take into account other variables, such as inclement weather and how it may impact daily water pumping. The amount of energy that can be produced by your home's solar power system depends critically on the orientation of the roof that houses the panels.

The storage tank will get a vertical push of water from the pump. The friction loss caused by the water moving through the pipes will be felt. The quantity of energy necessary to provide your household's daily water needs depends on the friction head loss. You can use helpful tables to determine the precise amount of solar panels you'll need to pump enough water to meet your needs.

You can choose the best submersible solar pump to satisfy your home's water needs if you have the correct information. There are several charts for various pumps to assist you in selecting the model that best fits your overall water needs. Other pumps have higher hourly pumping rates based on things like the solar power system's wattage.

After determining the type of pump you can use to meet your water needs, you should think about how many watts of solar power you will need to power the pump. Use the table that often comes with the various pumps to determine the appropriate quantity of actual solar panels. As a general guideline, you must make sure that the size of the panel or the quantity of the panels is roughly 30% greater than the energy requirements. This will assist in making up for other factors like bad weather that may actually reduce the effectiveness of your solar power system. The amount of energy required to pump water sufficient for your daily needs depends on a number of additional criteria, such as the height of the storage tank.

The wattage of the solar panel you'll need depends on the voltage of the water pump. You will require a 116 watt solar panel to power a 30 volt pump. As you have previously noticed, in order to have enough energy during periods of bad weather, a pump's wattage needs to be increased by roughly 30%. The instruction instructions for the solar power system that you wish to utilize to pump water must be carefully read.

Required Solar Power for a Fridge

You must determine the energy needs for your electronics if you decide to use solar electricity in your home. For instance, unlike other equipment like ovens, a refrigerator does not consume a lot of power. You will need at least 600 amp-hours of power to run a household refrigerator for around 24 hours. To charge the batteries within, 600 watts of solar panels are also required on the roof. To ensure that you get the appropriate solar panels, you must be aware of their various wattages and sizes.

A domestic refrigerator may require two solar panels each of 300 watts or less. When the sun is shining brightly enough, these panels perform at their peak. Additionally, you need to purchase powerful batteries that can store enough energy for use at night. In this scenario, disregarding other gadgets like kitchen appliances, radio, computers, lights, and television, you will just need two solar panels for the fridge. To determine how much solar electricity you could need, you must determine the energy requirements for each gadget in your home.

If you don't have regular utility bills, you might try to figure out how much electricity each device in your house uses. Do the same for other home appliances after determining the amount of power needed for your refrigerator. Use the overall wattage of your devices as a starting point to calculate how many solar panels you would actually need. To keep your devices powered up at night or inclement weather, it is crucial to purchase storage batteries with the necessary power. If you are unsure about how to use a particular device, seek advice from an expert. Many solar energy firms have helpful brochures that provide this kind of information.

The other crucial point to keep in mind is to use more solar energy than you need in order to make up for any losses brought on by inclement weather and irregular electricity use. To protect your technology, you must first speak with an expert. To prevent harm to your electrical gadgets, ensure that the power in your home is regulated. Additionally, you must make sure that your solar power system is installed correctly if you want to benefit from the effective energy output.

Since it significantly reduces your carbon footprint, solar energy is growing in popularity. The long-term benefit of having a solar power system at your home outweighs the initial expense. Solar energy is economical since it reduces your need to pay monthly utility costs. You must initially make the appropriate choice if you want to take maximum advantage of solar energy. Make sure you purchase the appropriate amount of solar panels to satisfy your energy needs. This is dependent on how many devices you use at home.

The overall amount of energy you need in your home must be calculated using your prior electricity bills. To determine the amount of panels you would require, you should also take into account other aspects, such as the size of your roof. The size and wattage of each panel are crucial factors in figuring out how many you'll need. You must speak with a qualified solar installer who can evaluate your property and your energy needs if you want to make the best choices.

Installing Solar Panels

In order to install solar panels, you must be confident working at heights, on roofs, and with waterproofing to prevent water leaks. Installing hooks, rails, or battens accomplishes this. You might need to install an under-roof screen or a waterproofing system, depending on the solar panels and how they are attached. To handle the solar panels and intervene on a roof, good physical fitness is also required.

Basic actions

- deteriorate where the solar panel is located
- Install the side and lower covers.
- Install a waterproofing system or under-roof screen
- Place the panels and rails.
- The inverter must be connected to the grid.

Tools needed

- drill driver or screwdriver with circles or sockets
- Square saw
- Rule for power extension cords
- ladder or scaffolding
- ladder for roofers (based on coverage)
- Key bushings or flats
- Screws
- Fishplate and other rail-installation hardware
- Anomaly or zinc
- Lead foil
- Battens
- Utilizing a silicone glue extruder gun
- Depending on the arrangement, electrical cable with IRL tubes and/or bridges may be required.
- gloves for protection
- Anti-fall mechanism
- protective eyewear
- Safety footwear

Deterioration at the Solar Panel's Location

There are two separate instances. The roof will need to be partially removed if it is covered in mechanical tiles or slate. Delineating the specific spot where you want to install the solar panels is the first step. Additionally, a surface somewhat bigger than the surface of the panels to be laid must have the tiles removed. A hole in your roof appears once this is finished, and after careful planning, the photovoltaic panels will be housed there.

Be careful not to damage the half-posts or the ridge tiles if you are cleaning up close to the ridge. If your roof is not planked (has a truss construction), you run the risk of falling if you walk on the under-roof screen or in the midst of the battens.

Set up the side and lower coverings

The arguments serve as connecting threads between your roof's roofing materials and the numerous protruding elements (such as the solar panels and their fasteners). Each type of mechanical tile has modifications, both in terms of shape and color. It is a sheet in the hue of slate or an antique piece of zinc in the case of slates.

The installation entails screwing a fastening onto the existing or newly installed battens. To record the precise measurements, a screwdriver and meter rule are required. Screws are also used to secure the hooks.

Install a waterproofing system or the under-roof screen

The roofing screen has been installed

A roof installation requires the installation of an under-roof screen. The latter is a technical plastic film with the purpose of forming an airtight barrier that is impenetrable to water. It is advised to leave an air gap between the under-roof screen and the insulation complex when installing the screen. If this is not possible and the problem lies with the insulation, pick a screen with a high water vapor or HPV permeability. This permits vapor to move and lowers the likelihood of condensation in the insulating complex. The insulation loses effectiveness if humidity forms inside of it, and it could even start to decay. The manufacturer's specified adhesives must really be used to cover the strips, per their instructions.

Putting in place the sealing system

The solar panels may be supported by a waterproofing system made of flexible plastic depending on the installation. This will be erected on the roof, necessitating the use of a solid foundation built of battens. Then, at the location of the panels, these sealing systems are attached onto the battens.

Install the Panels and Rails

Using hooks, the rails must be firmly fastened to the rafters (use a screwdriver and the proper screws for this). The solar panels are substantial and provide excellent wind catching. Complex shapes that can adapt their shape to specific tiles allow for attaching through the cover with possible alterations. Since the rails' structure is above the cover, full derailing is no longer required.

There are fishplates and numerous more connection components available if the rails need to be extended or crossed. It should be noted that some solar panels can rotate up to a quarter of a revolution with their long side vertical. The rails may then also need to be laid vertically.

After that, hooks or bolts are used to secure the panels to the rails. The upper cover can be fixed when all the panels have been placed precisely, and the missing tiles or slates can then be added. The support rails can be mounted directly on the waterproofing system when placing solar panels on it.

The Panels must be connected to the Inverter

An electronic gadget known as an inverter actually transforms the erratic direct current. This inverter is linked to the panels and the non-consumption meter on opposite sides of the device. At the top of each individual electricity generating installation that is connected to the network is this meter. For this stage to be completed correctly, understanding how to read an electrical diagram is essential.

It is necessary to connect some panels in series and others in parallel.

When two panels are connected in series, a voltage of 24 V is produced if each panel generates a value of 12 V. The panels are connected in parallel if the terminals are connected with the same polarity. The voltage this assembly produces is 12 V, however the intensity it produces is the total of the intensities from each of the panels. The tensions of each group must be equal if two groups of panels are connected in series or parallel, though, in order to protect the panels and increase their output.

DIY Solar Power System

If you look around your town or city, you'll probably notice that the number of solar panels is rising. When it comes to the rising cost of power, many people ask themselves, "Should I invest in solar energy?" Living off-grid while yet having a reliable power supply is gaining popularity. You can potentially save a sizable sum of money over the course of a year by using a solar power system to provide you with clean, renewable electricity. Solar-powered devices use a clean, renewable energy source that emits no greenhouse gases. This type of energy is known as green energy due to its benefits to the environment and the economy.

What Is a Solar Energy System Exactly?

A photovoltaic system is a collection of solar panels that converts solar energy into electrical power. Photovoltaic (PV) technology is used in sun-harvesting devices like solar panels to transform solar energy into an electric current. Solar energy generates energy that may be utilized for a variety of things, such as cooling, heating, and lighting up buildings. You might have enough energy from a single solar panel to recharge your batteries and supply all of your lighting requirements. As an alternative, you might set up a sizable collection of solar panels that would provide enough electricity to run your entire home. In order to supplement their current energy sources, the vast majority of homeowners in industrialized countries have installed or plan to install solar power systems in their homes. Even in developing nations like India and China, solar energy is regarded as a practical answer to the world's impending energy crisis. Our tutorial will walk you through every step of setting up a DIY solar energy system when it comes to designing and installing a solar setup in your home. As a consequence, you will be able to supplement your energy sources and also have a powerful solar power system that will meet your energy needs. Are you really interested in learning how to do it? Kindly keep reading!

What Functions Does the Solar Power System Carry Out?

The earth's surface reflects thousands of small solar energy packets (photons), producing enough solar energy to provide a sizable portion of the world's electricity needs. When photons hit solar cells in a solar panel, they release free electrons that are what produce electricity. Photovoltaic (PV) technology is used in solar energy systems to capture solar radiation and transform it into a form that can be used by the user. Numerous cell arrangements are employed to build the solar panels that generate electricity. The semiconductor material silicon, which is used to make the cells, absorbs solar energy. As a result, electrically charged particles begin to form and move inside the PV cell in response to the internal electric field, turning solar energy into electricity. In photovoltaic cells, silicon is structured into two layers: a positively charged top layer and a negatively charged bottom layer. Direct current (DC) is produced when sunlight reaches the cells because of the electric field that exists between the two layers of the cell. If you'd like, you can employ this current inside your home's outside walls. Although direct sunlight is not always necessary for solar panels to perform well, the amount of power that can be produced depends on the brightness of the light. If the amount of electricity generated is greater than what you need, you can buy batteries to store the extra energy for use at night or on overcast days.

Different Solar Power System Types

There are three different ways to integrate solar power systems into residences and electrical grids:

- Grid-Tied
- Off-Grid
- Hybrid

Solar Power System Linked to the Grid

To produce and distribute solar energy, a grid-connected solar energy system must be connected to a network or a utility power grid. It is significant to highlight that the system is not powered by batteries and that any additional energy produced is fed back into the network. Let's say your switchboard produces more solar energy than your residence consumes in terms of electricity. In that situation, the surplus electricity is fed into the system via the meter. The quantity of kilowatt-hours (KWh) of power that are sent to the grid is recorded by this specific meter. Every billing cycle, your electric company will track this number and send you the necessary payment.

Solar Power System Off-Grid

A link to the electrical grid is not necessary for an off-grid system to operate. The battery must only be connected in order to store the excess energy. Because of the battery capacity, it is feasible to preserve enough electricity to be used through the night and on overcast days. The electricity generated can meet the needs of residential, industrial, and commercial users even in the winter.

Solar hybrid power system

A hybrid system combines off-grid and grid-connected components. They function as on-grid solar systems when the grid is reachable and as off-grid solar systems when it is not. They generate energy on par with conventional grid-connected solar systems. As a backup power source, the associated lithium-ion or lead-acid battery is used. A hybrid PV system should be taken into consideration if you want to make sure you have a constant power source in the event of a power outage.

How Long Does It Take for Grid-Tied and Hybrid Systems to Pay Off?

You might anticipate a decrease in your utility company's monthly power cost when you switch to solar energy. Consider that your solar power system is wired to the grid. If so, you can sell any extra solar energy to the utility company and get paid on a quarterly basis based on how much power you send to them. If your solar panels aren't producing enough electricity, you can import more fuel from the utility. This will lead to a decrease in the cost of your electricity when compared to when you are solely dependent on the utility supplier. If you want to reduce your costs while also saving the environment, investing in solar energy is a wise decision. Even while installing a solar system has a large up-front expense, it will eventually pay for itself. The time it takes to pay off the loan is shortened by state and federal tax advantages.

A grid-connected solar system will typically pay for itself within three to six years if you build it yourself; between five to nine years if you use a professional builder. Aside from that, solar panels have a 25-year warranty, and any energy generated after the initial payback period is counted as an investment return (ROI). The payback period is the length of time it will take you to recover your initial investment in a solar energy system. Using an online solar ROI calculator to determine your payback period is easy and uncomplicated. To determine how long it will take for your investment in solar energy to pay for itself, you may really utilize the solar payback technique. Start by calculating the total cost of installing a solar system, taking into account any potential subsidies. Once you know how much it is, you may compare it to the price of electricity from your neighborhood utility company.

Payback Period is calculated as follows: (Total Installation Cost - Incentives Value) / (Cost of Electricity per Unit * Annual Electricity Consumption).

As part of your overall installation costs, you should budget for the price of solar equipment, a permit, installation charges, contractor compensation, and any other costs associated with the project. Rewards for incentives have a monetary value, such as the tax credits you receive for installing solar power. You are qualified to receive the solar investment tax credit, which is typically worth 26% of your federal taxes. Consider any additional municipal or state incentives offered in your country when calculating the repayment period. The billing rate per kilowatt-hour of energy consumed is used to compute the cost of electricity. You should be able to get information on fuel prices from your utility provider. Power usage is frequently shown on your electricity bill. Calculating the yearly usage of power requires multiplying monthly use by twelve. As an alternative, you can calculate the appropriate amount by gathering monthly utility invoices from the utility supplier for the entire year.

Solar power's costs are decreasing

The price of energy has fallen as a result of the development of affordable and effective power systems. The International Renewable Energy Agency (IRENA) 2019 report found that while global energy prices have declined by 82 percent, the cost of photovoltaic solar energy has decreased by 47 percent. The cost reduction was a result of advances in technology, economies of scale, and supply chain competitiveness. The International Renewable Energy Agency (IRENA) reports that between 2010 and 2019, this boosted global solar capacity from 40 GW (gigawatts) to 580 GW. The cost of photovoltaic (PV) projects that use renewable and ecologically friendly electricity sources has drastically decreased recently.

Installing a Solar Power System Yourself

Reduced installation costs are mostly due to the growing use of solar energy in residential areas. To keep the installation prices within your spending limits, use the do-it-yourself route. Do you therefore believe that you will be able to create a solar energy system on your own? Yes. While installing a solar system on your own can save you money in the United States, if you live in Australia, for instance, you must work with a licensed electrician and go by all relevant laws and regulations. Although installing a solar system may seem like a challenging endeavor, it is not impossible! To do the job on your own, all you really need is a basic understanding of electrical theory. However, there are a number of other factors to take into account before installing a solar power system in your home. To start, you must ascertain your annual solar energy usage as well as your electricity needs. The electrical needs of various housing styles differ. However, in some homes, energy is required to power a television, refrigerator, and other appliances, among other things. In some families, electricity is just needed to operate lights and charge cell phones.

A Few Things to Think About Before Building a PV System

The following things need to be taken into account before buying and installing any solar-powered equipment: 1. Due to the substantial financial commitment that installing a solar system entails, you must be precise in your calculations and decision-making. The proper products for your solar installation will be much easier to choose if you are aware of these crucial factors. Some of these factors to think about are as follows:

How long do you genuinely intend to stay in your home?

Even if your electricity prices drop right away after installing a solar power system, it will still take at least five to six years for your solar energy investment to pay off. Your investment in solar energy will be well worth it if you plan to stay in the same home for the foreseeable future; otherwise, the cost of solar energy for your home would be prohibitively high.

How much energy do you actually need to power your home?

You should also take into account how much electricity is needed to run all of the gadgets in your home. It should go without saying that different locations have different power needs depending on the type of equipment they use and the number of occupants in that particular home. For people who use less electricity, fewer solar panels can be required. As a result, figuring out how many solar panels will be enough to meet your energy needs depends on understanding how much power you typically use. The insolation (sun exposure) and the amount of energy you utilize, in general, decide the size of your solar power system.

Solar system price

Due to the fact that solar equipment might vary from one manufacturer to another, it can be challenging to assess whether it is appropriate and not costly. Although the cost of solar panels has continually decreased, the installation and related equipment costs can be too high. Taking everything into account, it can take you five to six years to make back the money you spent installing a solar energy system in your home.

Roof design

Solar panels may be installed on almost any rooftop, although some need more effort and cost more money to install properly. Installing solar panels on conventional wood shingle roofs is quite difficult. If your roof is of this type, you should consider hiring a professional with experience in rooftop installation services...

Location

The position of your house and the amount of direct sunshine your panels receive affect how much electricity they can produce. The amount of sunlight that reaches your solar panels may be diminished if you live in a neighborhood with many tall trees and structures. In the United States, if you reside in a state like California, Texas, or Arizona, you may experience a significant increase in the energy output from your solar system because these regions receive full sunlight for prolonged periods; however, if you reside in a state like Montana or Minnesota, you may still be dependent on the grid while utilizing solar energy to the fullest extent possible by outfitting a more efficient PV system with additional solar panels.

Permits

Before installing solar panels on your home, you must obtain all relevant licenses, including those for electrical work, building, and a specific solar PV permit. Consult a local expert to make sure you are aware of all applicable laws and zoning restrictions in your country. In order to avoid having to redo the entire system installation or suffering consequences for failing to do so, you need also obtain the necessary information on construction permits. In fact, this is especially true for solar panels that are buried in the ground. At the local level, permits for construction are frequently issued. Therefore, when applying for one, it is crucial to abide by all state laws and regulations that are relevant to your town.

How to Install a PV System on Your Own

- Your solar power system should be designed and sized based on your energy needs.
- Buy solar-powered devices.
- A reasonable option is to put in solar panels in your yard or on your roof.
- Connect the charge controller and battery (in battery-backed systems) to your solar panels.
- Connect your inverter and battery to the grid and install solar inverters and battery stands.

- For grid-tied and hybrid systems, consider installing a smart meter or using net metering to save money.

- The last step in finishing it is connecting your power system to the main electrical board.

- To connect your approach to the grid (if necessary), speak with your local energy provider about acquiring Permission to Operate (PTO) under the net metering scheme.

The steps given in this guide will help you set up your solar power equipment. The financial advantages can be attained over time because solar energy is a cost-efficient and renewable source of electricity. If you're interested in learning more about how to set up your own system, keep reading.

Setting Up Your Solar Grid Network

How to Create Your Own Solar Grid System

Step 1: Create a solar system that genuinely meets your needs as your first step.

First and foremost, you truly need to understand how much energy you require and how your energy usage changes throughout the day and year. To effectively build a solar system, you must have a firm grasp of the daily and seasonal patterns of your household's energy demand.

Step 2: Determine the ideal location for your solar panels.

Positioning your solar panels strategically in a location that avoids excessive shading, enables passive cooling, and allows for easy access can increase their efficiency during their lifetime. In this case, solar panels designed expressly for usage in low-light conditions will be effective.

Step 3: Order the right parts for your solar system in step three.

Once you have chosen where to place your panels, you will actually have a greater grasp of the power potential of your site. As a result, you will have the freedom to pick and buy the right solar system parts.

There are many choices available to you, including the following:

- The size and model (PWM or MPPT) of your charge controller
- The dimensions and quantity of your solar cells.
- The battery bank's capacity and the kind of batteries it contains
- Choosing whether the loads should be DC or AC and the overall voltage of each leg
- if there is one, the inverter rating;

Step 4: Build the specific housing or compartment for your solar batteries.

The next thing you must do is set up the battery house after placing your order for the individual parts. It might be a space within the home you now occupy, a detached storage building, or even a portion of the garage.

It is your responsibility to ensure that anything or anyone that could endanger the battery is kept out of the room where the battery is kept. Additionally, this might stop the release of the acids that are there.

Please be aware that most batteries do not perform well in cold conditions and so require some form of temperature control.

Because they are more cheap, unsealed batteries must have adequate ventilation in the battery compartment to prevent the development of potentially harmful hydrogen gas. This is because just a tiny amount of the dangerous hydrogen gas is released during the charging process for these types of batteries.

In the great majority of solar system installations, critical power electronics including inverters, charge controllers, safety fuses, shutoffs, and breakers are frequently housed within the battery compartment to reduce costs.

Step 5: Put the solar panels together.

The following step is to construct the support framework for the solar panels and put the solar array together. When kept rigid, properly cooled, and pointed directly towards the sun, solar panels last much longer and provide more energy. When installing solar panels, do bear this in mind.

You can build a solar support system in a variety of methods, depending on your available resources and, of course, your level of ability. The most efficient approach is to build a metal or wood A-frame structure with a south-facing orientation.

Additionally, it's helpful to have the option to manually adjust the angle of your solar panels in the winter and summer. In this way, your power output might be enhanced by as much as 40% with hardly any additional cost. Another approach is to build a one- or two-axis tracking device out of common home items.

Step 6: Installing your off-grid solar system.

Running the electrical cabling will be the next step in assembling your solar power system after the installation of your solar panels. The good news is that performing this surgery need not be as challenging as it may initially seem. In comparison to a solar system that is truly connected to the grid, the complexity of your home's electrical system may be greatly reduced if you desire to live off the grid.

Installing an all-DC system is an option if you ultimately decide to live off the grid because it may be simpler and more effective than using another sort of power source. Another option is to use an air conditioning system.

How to Build a Solar Panel Rack

The next step after getting solar panels is to mount them on a frame or rack.

Measuring

Actually, measuring your solar panels should be your first order of business. Before precise measurements can be achieved, it is required to actually place the mounting brackets on the solar panels. The rack's height should not match the entire height of the panel, but it must be high enough for the mounting brackets to attach to it.

Next, determine the panel's width with the mounting brackets attached. Depending on how long a piece of wood you were able to obtain, this dimension will indicate how many solar panels you can fit on a single rack.

For my project, the longest 2 by 4 I had was just a touch under 8 feet long. As a result, I was able to build a rack that can support three solar panels. I built two racks specifically for the solar panels because the system being considered would have six of them.

Construction of frames

It should be noted that the outer boards of the box frame are turned on edge, while the inner runners of the box are flat and flush with the top.

The first step in building the frame is to put together a straightforward box. The measurement from the solar panel's bottom to the very top of the higher bracket, which is affixed to the solar panel, will be used to calculate the height. The number of solar panels that will actually be mounted to the rack will be determined by the length of the longest piece of lumber you have access to, and the width will be calculated by multiplying that length by the distance between the mounting brackets on each side of the solar panel.

The inner runners, which connect solar panels when they are stacked on top of one another, must then be cut and attached to the box frame. These runners are used to connect two solar panels that are butted up against one another. Using your measuring tape, determine the distance between the top and bottom boards of the real box. Then, cut a piece of 2 by 4 to that distance.

At each point where two panels converge, place one of these boards. These boards can be attached once they have been turned over and are flattened, flush with the top border of the box frame. This provides enough space for mounting the two brackets that are included with each adjacent panel.

The next phase is to build the rack system's framework. You'll need to utilize the angle finder at this point. The ideal angle at which you should build the rack will depend on where you are on the globe.

You must add 15 to the Latitude for the winter angle and deduct 15 from the Latitude for the summer angle if you are especially designing a rack system that can be tilted up and down for the maximum effectiveness of your specific panels from winter to summer. You can then choose the ideal angle for your rack system using this information.

The angle finder should be positioned so that it is perched on one of the side boards of the box frame, which should be set down on the ground. The top border of the box frame should then be raised until it is at the correct angle. Now that the top of the box frame has been hung at this angle, you will need to take two measurements.

It is not always required to identify the exact angle that is most effective.

The vertical drop from the top of the box frame to the bottom of the base is the first measurement. These will act as the two braces that are attached to the base's ends.

The vertical distance that must be spanned between the bottom of the box frame resting on the ground and the place where the braces will meet it on the ground is the second measurement. Two boards that have been cut to those measurements are required. By attaching the two end boards to the longboard, which is the length of the rack plus 3 inches, you can now start constructing the foundation.

You are now in a position to attach the box frame to the base, and it will already be at the angle necessary for a static solar panel rack system to function as efficiently as possible.

You have finished the task once the panels are fastened to the rack.

Creating a Solar Grid Network Wiring

The panels must be permanently connected to one another in order to create an electrical circuit through which electricity can flow in order for a solar photovoltaic system to be operational. For the DC power generated by the panels to be turned into AC power that can actually be utilized in your home and supplied to the grid, the panels also need to be linked to an inverter. pertaining to solar energy. Each group of panels that is connected together is known as a string, and this procedure is frequently referred to as "stringing."

Parallel versus series stringing

Solar panel wiring can be carried out in a variety of ways. One of the most crucial distinctions to understand is the one between connecting solar panels in series and parallel. Each of these various stringing configurations has a distinct impact on the circuit's current and voltage levels.

Series-connection of solar panels

The left side of the diagram above shows how to connect solar panels in series by lining up each panel in a line and attaching it to the panel in front of it.

Similar to the common batteries you are undoubtedly already familiar with, solar panels include positive and negative terminals. When solar panels are connected in series, the wire originating from each panel's positive terminal is connected to the next panel's negative terminal, and so on.

With each additional panel added to a series of panels, the total voltage (V) of the string rises, but the total current (I) flowing through the string remains constant.

One disadvantage of connecting components in series is that a darkened panel may reduce the amount of current that travels through the entire string. The total current is reduced until it reaches the level of the panel that draws the least current since the current remains constant along the string.

Parallel solar panel connection

Installing a solar array with parallel-connected panels, as seen on the right side of the image above, is a little more challenging. When actually stringing solar panels in parallel, all of the positive terminals on the string are linked to one wire, and all of the negative terminals are connected to another wire, as opposed to connecting the positive terminal of one solar panel to the negative terminal of the next. As a result, it is not necessary to connect the positive terminal of one actual panel to the negative terminal of another.

When panels are connected in parallel, the current (amperage) that flows through the circuit increases with each additional panel while the voltage (equal to the voltage of each panel) remains constant. As a result, one advantage of stringing in parallel is that even if one panel is being overly shadowed, the other panels may still function as intended, and the current provided by the entire string won't be impacted.

Maintenance of an Off-Grid Solar Power System

You have to learn how to maintain the PV system on your own while you are separated from everything and reliant on it, let's look at how you can do it.

PV Maintenance Styles

Corrective and preventative maintenance are the two forms of maintenance available for PV systems. When the system malfunctions comes first. It typically manifests as a loss of power or a lack of power production for a variety of reasons, in which case the system needs to be addressed (components, parts, etc. replacement). Corrective maintenance is what it is known as. It frequently occurs as a result of neglecting preventive maintenance, which is intended to check the state of the system's components to ensure everything is in working order.

Each user determines how often this maintenance is performed. However, I advise performing preventive maintenance on the PV system at least once a year or every six months. Furthermore, you need to be an expert on installation.

WARNING: Preventive maintenance won't stop your system from breaking down, which can happen if your panels and inverters were installed improperly or are of poor quality. Buy the best ones afterwards if you want to prevent major issues.

Bunker Panel Installation Considerations

What is preventive maintenance used for?

The major task of maintenance is to search for any environmental elements, such as dust, water, moisture, humidity, filth, etc., that may have an impact on the system's performance.

The following tasks are carried out by a reputable maintenance service:

- Watch out for all cable ties, conduits, and power wires.
- Junction boxes and safety devices should be inspected and cleaned.
- Solar panels should be cleaned.
- Clean the interior and secure inverter connections once more.

As you can see, there may be a significant buildup of dust in the inverter cooling channels. In this situation, preventive maintenance can get rid of all the dust and make the inverter fans work at their best. In the event that this is not done, dust may obstruct the cooling system, lower inverter efficiency, or even harm the inverter itself.

If you want your solar system to last more than 25 years, you must follow the manufacturers' recommended schedule for preventive maintenance. Keep in mind that the entire system is evaluated; it is not simply the surfaces that are cleaned. The cost of maintenance is modest in comparison to the operational issues that could develop if it is neglected, not to mention the possibility of it breaking down while you are stranded in the bunker without access to repair services or replacement components. Remember that every six months is the optimum frequency for maintenance, and ideally a prepper would anticipate a system lasting more than 25 years.

Upkeep of various PV components

Solar photovoltaic panels are low maintenance, however installing them on flawless surfaces will increase their performance and lengthen their usable life.

Because of this, I am now committed to educating you on how to maintain and clean solar panels in detail.

Regular cleaning can work wonders and improve the efficiency of your equipment because leaves, dust, pollen, filth, and even genuine bird droppings can impair energy production.

How Often Should You Begin Cleaning Solar Panels?

Don't forget to keep the system running.

Only a complete shutdown of the gadget is possible if you need to make any repairs.

Solar panel washing doesn't require elaborate procedures or expensive cleaning agents. To remove accumulated dirt and debris, just use mild soap and water.

If the dirt is more tenacious, a soft-material scraper or sponge with an extension handle is needed to get to tight spaces without scratching the panel surface.

If you want to know which cleaning products or materials should or shouldn't be used, it is recommended to refer to the manufacturer's instructions or operator's manual.

Even though the operation is straightforward, it is best to clean carefully in order to spot damage or assess the status of the PV system.

whenever possible, use a hose, and aim the water upward. Try to avoid getting water on the panel directly, while it's not a big deal if it happens. Additionally, safety must always come first before cleaning or beginning any job, and you must be prepared with all essential tools (such as seat belts, ladders, footwear, etc.) to access high locations and climb on roofs.

When Should Solar Panels Be Cleaned?

Solar panel cleaning is best done in the late spring and late fall. Pick an hour of the day when the weather is calm and pleasant. Consider it like a shot glass: cold water can make panels more prone to shattering from abrupt temperature changes. The water you use may also evaporate fast and produce scuff marks if the sun is fully out and the surface is exposed to a lot of radiation and heat.

Cleaning solar panels first thing in the morning is very beneficial because the spray from an overnight installation will assist remove accumulated dirt, using less water and less energy to clean. The plate needs regular upkeep.

The best times to actually clean your solar panels are also on cloudy days or in the evening. Imagine you reside in a region that experiences extreme aridity, is rural, or has significant levels of air pollution, such as one that is impacted by neighboring buildings or an industrial sector. If so, you might need to perform more frequent maintenance on your installation. But it should only happen every three to four months at most.

In general, it is advised to clean about three or four times a year, but in locations where solar panels are exposed to more light, such as rural areas or locations near the sea, more frequent and thorough maintenance is advised, as well as more frequent inspections of the condition of photovoltaic and solar installations to identify potential system failures. This should also be taken into account if the panel's surface is more inclined and horizontal.

Visual Examination of Solar Panels

- Verify the support panel's construction. Typically, the panels are made of aluminum with corrosion-resistant stainless steel parts. However, it is crucial to check the cover for irregularities like cracks or stagnation on the board's surface, as well as to see if it is securely fastened.
- Verify the electronic parts. Keep an eye on inverters, DC panels, AC power, and other devices and systems.

- Verify the storage mechanism. You should think about keeping the top surfaces of the cells, terminals, and connecting terminals clean in isolated installations, i.e., installations with storage systems (solar cells). Additionally, when the battery requires repair, more specifically:

- To extend the battery's life and ensure good performance, recharge it as directed by the manufacturer, if necessary.

- Check the battery's capacity and state of charge with a hydrometer by gauging the electrolyte density. You might not have this equipment; if you are unable to conduct this measurement, try calling a qualified specialist.

- Equalize the battery on a regular basis to maintain control over its storage capacity, boost performance, and lengthen its lifespan.

Additional Solar Panel Upkeep Jobs

You can think about employing a pressured hose system to clean your solar panels if your installation is in an inhospitable spot or at an awkward height. It is not suggested that you do this. It does more harm than good since the force of the pressurized water can damage the panel's surface. Additionally, avoid attempting to access a high-rise installation without the necessary safety gear.

When water is involved, the risk of working on a roof rises, making any task more hazardous and slippery. For your protection, it is best to actually hire a qualified specialist or a professional business.

Your stand-alone installation may be less effective if your panels are covered in accumulated dust or debris. Saline surroundings also have an impact on panels that are near the ocean.

Advice on How to Care for and Maintain Solar Power Over Time

Remember these general pointers:

- Only the top of the module should be washed with water.

- Use a mild detergent if you must use something other than water. Avoid using abrasive, harsh cleansers.

- Wipers for the windshield might help with cleaning. Once more, avoid wearing anything that could scratch the panels.

- To prevent a thermal shock, clean the panels in the morning or evening when they are cool.

- With a soft cloth or a windshield wiper, you may dry them.

- Prevent the solar panels from being shaded by trees and other vegetation. It lessens their output.

- Ask an expert for assistance on solar energy matters: If you decide to install a solar energy system, look for a qualified installer. By doing this, they will conduct the appropriate research for your particular situation, and you will have experts by your side who can constantly offer you advice in the future if necessary.

In terms of repetition

When deciding how frequently to clean and maintain solar panels, the weather and their look will be important factors to take into account. It always depends on the circumstances they confront and the area of the country where the installation is located, but the frequency might range from once to twice a year.

The Solar Kit Spaces' Step-by-Step Inspection

Solar inverters are unique components that require careful inspection at each stage. It's crucial to remember that they shouldn't be placed in an area where they won't be subjected to extremely high temperatures (over 35°).

- Every year, check the wiring because it will eventually deteriorate.

- Check to see if the voltage and current are accurate.

- Examine the operating error signals and alerts. To do this, you must consult the manuals for that model.

Solar Cells

You must perform a variety of upkeep and care procedures if your solar panel package includes batteries.

- At least once every year, assess for damage and perform exterior cleaning.

- Check for potentially damaged places, among other things, and make sure connections are secure in addition to cleaning.

- Four to five times per year, make sure the water level is accurate.

In order to guarantee the longevity of the photovoltaic kit, regular maintenance of the complete installation is necessary. In addition, by ensuring optimal performance, you will ensure that the solar panel kit will produce more energy.

Maintaining and Monitoring Your Home Solar Power System

Technical and performance concerns are common, and you may quickly identify faults by using a solar monitor or the information the inverter displays. You may either "see" how your system is doing or remotely modify your consumption to keep your electricity usage in control during times of high demand, depending on the solar monitor you choose to install. If significant variations in energy production are detected, certain monitoring systems may send warnings to your mobile device.

Here is a list of the things a top-notch solar monitoring system can achieve for you.
Monitoring at the production level: This is the lowest level of monitoring. You can always find out how much electricity your solar panels are producing thanks to your monitoring equipment. The production readings can be seen using an app that comes with your monitoring device. This information is represented digitally.

Additionally, there are several gadgets available, with prices ranging from $20 to over $300. One of the top manufacturers that is simple to install is Sense Energy Monitor. The device may be installed in less than an hour, but there are worries that it may not always provide reliable data. High-end solar monitors from SolarEdge also have defect detection. Because the monitoring system is built within the inverter, you must buy a SolarEdge inverter to use this system.

Monitoring of consumption: You may keep track of how much electricity you use, and over time, your device will figure out which appliances use the most. You'll have control over which appliances run at what time of day thanks to this. In order to give solar panels enough time to charge the batteries if they are installed, you should often use appliances that need more electricity during the peak sun hours.

This tool is fantastic if you sell electricity to the utility provider. Bill comparison. It is simple to compare the cost of your electricity to the amount that the utility company pays you for the power that you export to them. The utility provider can also provide you with daily projections of how much you can expect to spend for the electricity you consume. This tool is helpful and effective for determining whether your solar system is helping you save money.

Troubleshooting

The two most frequent troubles that a residential solar system may run into are zero power output (no electricity) or problems with high and low voltage. The user guide that comes with your inverter should be your first port of call if you run into either of these problems. The majority of digital inverters contain a screen that displays trouble codes. The following steps can be used to resolve the issue if you are unable to pinpoint its root cause.

Zero power output: When a residential solar system has been working for some time, this issue is frequently encountered. Either an inefficient inverter or charge controller is at blame. Additionally, there is a chance that one of the panels could have failed. If your solar panels are wired in series, an open circuit will be created if one of them fails. Use a multimeter to inspect each component in order to repair the system. For a current/voltage test, you must choose your multimeter. A faulty panel is indicated by zero voltage or zero amps coming from the array.

High power output: The system may occasionally supply too much electricity, which could harm the appliances. This typically indicates that your inverter's internal voltage regulator is malfunctioning. If over-voltage situations are found, inverters should shut down.

Low power output: This happens when a solar system does not produce as much power as it is capable of. Fixing this issue entails more steps than dealing with a power outage. Among the causes of a lower power issue are:

- **The main common reason of low voltage is shading.** It is advised to regularly examine the installation site to make sure there are no nearby trees that would cast shadows at any time of day. The solar panels won't receive direct sunlight because of the shadow. While trees continue to grow over time, solar panels have a long lifespan. The canopy of trees varies depending on the kind of tree and how much shadow it casts.

- **Temperature**: As the temperature rises, the voltage drops. The system's ability to output electricity is reduced by high temperatures. Because you will increase the system voltage, adding more modules in series will solve this issue. Make sure there is enough airflow around the inverter and underneath the panels. On the inverter, make sure no apertures are blocked.

- **A faulty or loose connection should be investigated**. To identify the starting site of the low voltage, voltage levels are checked using a multimeter at numerous locations. The main root of this issue is poor wiring. Rewiring and installation should be done by qualified personnel, it is advised.

- **Solar panel flaws:** Despite a solar panel's long lifespan, failure is nevertheless common. Because product recalls happen frequently, you should pick a renowned brand. If you purchase inexpensive goods, it's likely that you won't learn about any flaws, and frequently, warranties are either nonexistent or nullified if you install the system yourself. When troubleshooting, you can look for the following on your panels.

- **Battery not adequately charged:** This occurs frequently with off-grid systems, especially if the system is used for a prolonged period of time. Batteries of lower quality cannot be guaranteed to have a certain number of cycles before losing their capacity to hold a charge. Use your multimeter to verify that the battery is keeping charge properly in this specific instance. If you notice your batteries are below 12 V after charging them, it's definitely time to replace the batteries. Each cell must report a voltage more than 12.6 V. Sometimes, there may be a problem with the charge controller as well. If your inverter includes a built-in charge controller, you should have the manufacturer inspect it.

Hotspots

A hotspot develops when the cells start to show microcracks. The solar cells in a panel will experience a succession of resistances as a result, which may have grown over time. As a result, the voltage in the panel will be decreased overall, lowering the voltage of the solar array as a whole.

Error in the Panel Junction Box

A junction box that has been exposed to moisture could develop more resistance. Because of this, it's important to make sure the local junction box you place behind the solar panel has a tight seal. Before mounting the panels on the roof, make sure there are no air gaps and give the glue time to fully dry.

Delamination

Delamination refers to the separation of the bond between the top glass and the polymers at the rear of the panel. This issue is visible to the unaided eye, and moisture can enter the panel's electrical circuit, which could result in a current leak or a short if there is delamination.

Alternative Troubleshooting Techniques

Zero-power situations can happen if surges, malfunctions, and overloads trip the breaker switches. This prohibits solar energy from supplying power to the main panel box or from charging batteries. When you turn the switch on and off and nothing happens, you should do a continuity test; if nothing occurs, your breaker may be damaged.

Examine the solar panels for obstructions: Unexpected shadowing may result from recently fallen tree foliage, collected dust, debris, or pollen that may gather over time. To stop mice, birds, and rodents from constructing a house out of your solar panels, think about putting a pet guard.

Verify your solar monitor; the information it provides will help you understand how your system is doing. The data logging is useful since it allows you to view the system's historical performance and identify the exact moment when performance started to decline.

Maintaining Your System

Even while solar panels don't need maintenance like other power producing equipment, you might need to do so sometimes. Regular physical inspections are necessary to find anything that can affect how your system works. The following are some things you can do to better look after your home solar system.

- Any trees or vegetation that can cause your panels to shed should be removed and kept in continual trim. Although your neighbor might have towering trees, you might not. Talk to them about your choices for preventing shade from falling on your solar panels. If birds frequently visit your property, look for bird droppings on the panels and wipe them suitably with a lint-free cloth that won't scratch the glass. The panels may become opaque as a result of cleaning solvents, and scratches may scatter light, preventing it from effectively reaching the cells. To clean the surface of your panels, you may only use water.

- Avoid using other types of lighting to make your solar panels produce electricity. Your solar panels may become damaged by the light from fires and torches, voiding your warranty and costing you thousands of dollars. Make sure your array receives only direct sunshine.

- Don't let any dust get near your inverter. Dust is a good electrical conductor, so if it gets inside your inverter through the cooling system, it could cause a short-circuit that could cause a fire or an electrical explosion. Additionally, water contact with your inverter is not permitted. If moisture seeps into the inverter, not only will your system be harmed, but you also run the risk of electrocution and damage to your electrical appliances.

Don't put the batteries on the floor if they are part of your system and have been installed. If you do this, the temperature of the floor and a lack of ventilation could cause harm to the battery cells.

Top Errors People Commit When Installing Home Solar Systems

Inadequate system sizing: Most of the time, incorrect system sizing causes solar users to fall short of their expectations. Grid-tied consumers frequently overestimate their savings and fail to take into account how much energy their solar system generates in relation to their usage. Solar system users in off-grid installations may experience blackouts. Off-grid users frequently overlook seasonal output variations (for example, wintertime solar panel production is typically lower due to less sun hours). Before installation, carefully perform your calculations and take into account seasonal fluctuations to assist you choose the appropriate system size.

Lack of vision: Solar users typically base their choices for the solar system on their existing expectations. Customers are only permitted to consider their present load and available space for expansion. A solar system is a long-term investment, therefore it is important to take this into account when deciding which system is best for you and your family. The degradation of performance over time should also be taken into account. Depending on the quality of the brand, solar panels lose roughly 1% of their annual production. You have to consider this. In order to make up for the output losses from your solar panels, you should slightly oversize your system.

improper solar panel installation: solar panel installation calls for accuracy. Your panels will be exposed to the sun in different ways if they are not perfectly aligned or orientated, which will change the production of each individual panel. Solar optimizers can assist in solving this issue, but they are limited by the data they are provided. Power imbalances must be avoided at all costs. A different location should not be used to install solar panels from the same array. I am aware that some individuals choose to place linked strings on either side of the roof in an effort to catch the sun as it moves. Especially if the panels are connected in series, this is a horrible idea. The performance of your system will be determined by which panels produce the least amount of electricity. Installing a ground-based system that makes use of solar trackers is preferable if you want to take advantage of the sun's location throughout the day. Solar trackers are extremely expensive, but they can easily increase your peak-sun hours.

Focusing on installation costs: most owners make the mistake of going with the cheapest solar system to reduce the time it takes for their investment to pay off. The less expensive solar supply sources might not perform as well as anticipated. Another drawback of purchasing inexpensive goods is that doing so will extend the time it takes for a return on investment while also jeopardizing the energy production process. Never skimp on your solar system's quality. Avoid setting up a monitoring system: The inverter's ethernet connector is there for a reason. Since professionals rarely talk about solar monitoring, I do not see this as a DIYer's error. You must install a monitoring system if you want your home solar system to maintain its optimal performance over its lifetime.

Threats to your home solar system in the future

Your home solar system is susceptible to cybercrime due to remote monitoring via the Ethernet communication ports that the majority of inverters currently include. Smart-grids are the future of the industry, and your linked home could be a point of entry for criminal activity. Consider the ransomware incidents that made headlines in 2021. A hacker could bar you from using your home system to access electricity during a ransomware assault. Or they could steal some of your electricity or surreptitiously alter your solar monitor to do this. On a broader scale, this has happened before, and recent attacks on the nation's infrastructure have resulted in blackouts in some areas. DDoS assaults, also known as denial-of-service attacks, can happen if a hacker breaks into your home network and gains access to and control over your inverter. You must be aware of system vulnerabilities, and if you stay current with manufacturer-provided software upgrades, these vulnerabilities can be quickly rectified. The easiest approach to protect yourself from cybercrime is to rely on your supplier's security patches if you aren't particularly tech-savvy. To avoid system intrusions that can be difficult to detect, you must also ensure that your home network is secure.

Bonus Advice

Find a helpful licensed electrician because problems will inevitably occur if you build your system yourself. A preliminary inspection should be performed by an electrician before you call the inspectors to request permission to work. By doing this, you can really lessen the possibility that your system will be declared dangerous. Most likely, you will have to give the certified electrician a modest fee so they can check to see if everything was done properly.

It will be challenging for you to thoroughly evaluate the state of your roof on your own and give an engineering report that may be required when you submit your application to get permission to operate. **Take into consideration a professional initial site visit.** Your roof's ability to handle the whole weight of the solar panels you intend to install will be quantitatively demonstrated by the engineering report, which will also assist you in selecting the proper racking solution.

Early application filing will help you avoid delays in getting your request approved. These applications can now be submitted online, and the local council will give you an appointment time for the inspector to visit your property. It can take between 30 and 45 days to complete this.

Make sure all of your components are compatible with one another by purchasing them all from the same supplier. Purchasing your supplies from a single vendor will provide you access to competent personnel who can suggest the best things in their store, saving you the stress of poring over technical spec sheets.

The Challenges, Peculiarities, and Possible Issues of Off-Grid Solar Systems

For those who live in places without access to electricity or where the cost of generating electricity is very high, installing an off-grid photovoltaic power generation system may be a solution to the problem of generating electricity.

The off-grid system can be more difficult to use than a typical grid-tied setup. You must be aware of the quirks, potential problems, and how to resolve them because such systems are intended to be the user's sole power source. The design of your off-grid poses the first of these difficulties. Off-grid units also have extra parts like battery storage and charge controllers in addition to the panels, power inverter, and power distribution system of a typical grid-tied system. All of them add to the complexity of the electrical system. However, since there won't be a backup from the grid to fall back on if the system fails, it must be built to be dependable.

Common Off-Grid Solar System Design Issues

If they are to withstand regular usage, they must be built with the user's energy needs in mind. When choosing the various system components, including the solar inverter, batteries, charge controllers, switch, and even the electrical components in the house or cabin where you wish to install the unit, this must be a top priority.

You must perform appropriate prior work before planning and designing the PV system. Understanding your load type and overall power needs is the first step in doing this. Additionally, you need to assess the climate at the location where the installation project will be done. You can only move further with designing your system to specification when this compilation has been properly completed. However, the following points should be noted when planning and carrying out the installation project:

Make that the battery's voltage and the voltage of the other components are compatible

Without any inductance equipment in the middle, the PWM charge controller and battery storage system are often connected directly by an electrical switch. As a result, your component voltage must be between 1.2 and 2.0 times the battery voltage. If your battery has a 24V capacity, for instance, the input voltage should be between 30 and 50V at most.

In addition to the switch, MPPT controllers typically incorporate an inductor circuit. If your battery is 24V, the input voltage should be between 30 and 90V. For these, the input voltage must be between 1.2 and 3.5 times the battery voltage.

Make that the solar energy system's output power corresponds to the controller's.

Take a 48V, 30A charge controller, for instance. It is recommended that the output power, which is often the product of these two factors, be 1440VA. As a result, the component's output should be kept at about 1500W. As a result, you should verify the battery voltage and multiply the output power by it when choosing your charge controller. In fact, this will help you determine the charge controller current that should be employed for the system.

You may employ several inverters

Inverters can be connected in parallel if the power of one is insufficient for the unit. In this scenario, the inverter output will have a different voltage, current, and amplitude because the PV system output will be connected to the load.

Common Off-Grid PV Solar Energy System Problems and Solutions

I've listed a few problems that you might run into when setting up or using your off-grid solar device in this session. Actually, the list of potential technical problems that you can encounter is not all-inclusive. However, you'll be prepared to act if you run across any of these problems.

The inverter LCD is not showing

It is likely that there is no current input from the battery if there is no indicator displayed because the inverter LCD is powered directly by the battery. This might possibly be the result of any of the following causes:

- **Low battery voltage:** Batteries are typically fully charged when they leave the factory. However, batteries that have been idle for a while will start self-discharging gradually. When the battery series connection cable is improperly attached to the input, a similar circumstance occurs. Battery charging is not possible, and the inverter cannot be powered by the available voltage.

The solar energy system cannot charge the battery when the voltage is too low. You must thus take it somewhere else to charge it to at least 30% before rejoining.

- **An incorrect connection between the positive and negative poles of the battery might result in this**. Typically, the black wire should be linked to the negative poles and the red cable to the positive poles. In the event that this connection is reversible, the battery will not function.

Normally, if the circuit has an issue, you should use a multimeter to measure the voltage on the batteries first. Each battery's voltage should add up to the overall voltage at a normal voltage. If this reading is inaccurate, you should double-check all of the terminal connections to be sure they are made correctly.

- **Another issue could be a DC switch that is broken or not connected**. It's also possible that the inverter is broken. When everything is in order with the battery voltage and terminal checks, you should examine the DC switch. When this is turned on and appears to be in good condition, the inverter may be at fault.

Not Charging Battery

Power from the PV module is passed through the controller to the batteries to charge them. It might alternatively go through the charge controller first, then the mains supply. The battery can not be charging for a number of different reasons. This might comprise:

1. Problems with the PV component: this could be caused by a low voltage in the PV component, a lack of sunlight, or a wiring issue between the PV component and the battery. To ensure that everything is connected correctly and that it functions, you should inspect and test the DC switch, terminal, PV component, and cable connector.

2. Bad battery circuit wiring: Verify that the battery circuit has been wired correctly.

3. Fully charged battery: The battery can no longer be recharged once it has been fully charged. An input overcurrent or input overvoltage could be the issue. A battery unit's charging current should typically range from 0.1 C to 0.2 C, with a maximum value of 0.3 C. The charging current should range between 20 and 40A and should not exceed 60A for a lead-acid battery with a capacity of 12V, 200A. You must make sure that the

power supplied by the PV component matches the charge controller's power requirements.

Input overvoltage may possibly be the issue. To make sure that the voltage from the PV component isn't too high, you should check the voltage of your battery panel. If so, you must lower the number of PV cells to align with the battery input.

The inverter displays "Overload" or won't turn on

The inverter, battery, or load could all be the source of an overload issue.

1. Inverter overload: The load needs to be adjusted if the inverter's power exceeds its time range.

2. Battery overload: in most cases, the discharge current should range from 0.2 to 0.3 C and not exceed 0.5 C. Therefore, the maximum output power shouldn't be more than 2400W for a 12V, 200AH LFA battery. Different battery models from various manufacturers may have different particular values for these parameters:

3. Overloaded loads: Elevators and other loads that produce a back electromotive force shouldn't be connected directly to the inverter's output terminal. The inverter will sustain damage from the generated reverse electromotive force. For this kind of connection, a frequency converter needs to be put in between the elevator motor and the inverter. Another possibility is that the inverter's starting inductive load is just too great.

Make sure that the voltage's rated power is less than the inverter's. Peak load power should typically not exceed 1.5 times the inverter's power.

Off-Grid Solar Energy Systems: Unique Features

Going fully off the grid is much more fraught with uncertainty than merely adding a grid-tied solar power system to supplement your energy needs. Before moving forward with an off-grid solar system, there are extra issues to take into account in addition to the technical ones.

Going Off-Grid Is Cost-Effective in the Long Run

Although the overall cost of the solar panels, inverters, and battery systems that make up an off-grid solar installation have steadily decreased, the initial investment is still not exactly cheap. In reality, the initial cost of constructing a fully off-grid system will intimidate the majority of customers. This expense also takes time to recover. Once you are off the grid, you won't have to pay your utility company, but it will take time for the accumulated cost savings to make up for the initial cost of installation. However, in most situations, it finally does, and constructing an off-grid system is a financially wise choice (provided all the other initial criteria you were supposed to evaluate check out). In most places, you may be able to save up to $20,000 over the course of 20 years.

Typically, five years after installation, you should be able to return your initial investment in the upfront cost of the solar panel itself. Even though it takes up to ten years, the solar energy storage batteries finally make back their investment.

Going entirely off the grid could be against the law in some places

Installing solar panels to produce power to supplement grid supplies is typically permitted. To entirely disconnect your house from the grid, nevertheless, may be against the law in several nations. Such limitations typically apply less to rural areas and more to urban or metropolitan places.

Technically, living off the grid is not fully prohibited in any US state. You are frequently permitted to engage in basic off-grid living activities and produce your own energy. However, entirely cutting off your connection to the traditional utility is heavily regulated in many locations.

Off-grid life is often governed by legislation that varies from state to state and may even differ from your particular county or municipality. Solar panel installation is often permitted. You will want a specific permit to do so in other locations. However, not all US states allow citizens to completely cut off from the electricity grid. While some states permit this, others still demand that homeowners pay the utility company a certain amount each month.

For example, even if you are not utilizing the service, you must maintain your connection to the grid and pay a minimal connection cost. In other circumstances, the law might not forbid disconnecting from the grid, but your insurance provider might insist on it in order to maintain your coverage. Therefore, be essential to research the legal nuances relevant to your particular locality before you start planning your off-grid installation. For further information, get in touch with your utility company or a reputable solar installation business in the area. Going off-grid is a logical decision in some situations, such as a mobile home or a secluded cottage in the woods, and there are virtually any limitations to such.

Implications for Real Estate

If you need to sell or rent your house, going fully off the grid may have significant repercussions. Most likely, an off-grid house will appeal to a variety of buyers. On the one hand, eco-conscious potential buyers will probably find the concept of a house with a small carbon imprint rather enticing. Naturally, this would result in a higher value, particularly if there are a lot of individuals in your area who are interested in off-grid life. However, many traditional home purchasers may not be overly enthused about the significant energy sacrifices that come with living in such homes. Be sure to consider both sides of this before committing to a project.

Taking Care of System Issues Will Be Slightly Difficult

Going entirely off-grid has repercussions, one of which is that you will be entirely in charge of fixing and maintaining your electrical system, as opposed to before when everything depended on the utility provider. However, the good news is that solar panels don't require much upkeep. It should only require occasional panel cleaning and periodic battery and inverter inspections to maintain the system operational.

But when there is a significant flaw in the system, things could become pretty difficult. If your electrical system suffers significant harm or a fault, you might need to hire a specialist, which is often rather expensive. And this might also imply that you won't be informed until the problem is actually resolved.

Off-Grid Solar Energy Needs Some Significant Lifestyle Modifications

Anyone who has a fully off-grid system will tell you that moving off-grid requires significant lifestyle adjustments and sacrifices. Off-grid living is more of a lifestyle than just a choice in reality. Off-grid solar is a no-brainer for people who live in places where connecting to the grid is challenging or impossible. In such locations, noisy, bulky generators with a potential price tag are the alternative.

For individuals acclimated to a simple way of life or environmental activists dedicated to reducing their carbon footprint, an off-grid system is also a simple option. These people already have a sustainable mindset, therefore it might not be too difficult for them to make the necessary lifestyle modifications to adopt off-grid living.

Off-grid solar power would not be practical for you, however, if you are a normal homeowner in a city or town who prefers to use unrestricted ordinary grid power. Going entirely off-grid might not be advised for individuals looking for a quick strategy to reduce their utility costs. This means that a grid-tied system will be a better option for you if immediate cost savings are your motivation for exploring solar electricity. They are less expensive than solar, and since you are still connected to the grid for backup power when you need it, you won't need to dramatically limit your electricity usage.

Homeowners on a tight budget would be better suited building a comparatively smaller system due to the high expense of an off-grid installation. Consequently, a significant lifestyle will be needed. You must reevaluate your energy usage and make some significant changes if you plan to live off the grid.

Weather patterns and yearly seasons may also place further restrictions on the daily electricity supply. For instance, you might only produce half as much power on a foggy day as you usually do. Running out of battery power might also put you in a difficult predicament because it requires you to wait for the sun to actually shine again or use backup generators in the interim.

Different facets of your life will be impacted by the restriction of having restricted access to power. Since you can't pump as regularly, you might need to conserve water, use air conditioning only when it's absolutely required, and get rid of large equipment that consume a lot of power but weren't part of your original plan.

Another alternative is to use more renewable energy sources, such as wind or hydropower, to enhance your current energy supply. However, compared to solar energy, these sources are frequently more expensive and less dependable. The adjustment to living off-grid might be made simpler by these additional possibilities. For instance, passive solar heating, which may be put alongside your PV solar cells, can take the place of conventional heaters. For space heating, you can alternatively choose traditional wood heating. A sustained transition to independent solar heating systems will be made possible by changes like these.

Legalities and Laws

Despite the fact that solar energy has been used for thousands of years, rooftop solar panels are still regarded as relatively new technologies. The fact that they are typically added to buildings after they are built makes them particularly challenging to adopt. Less problems are likely to occur when house features are considered in the infrastructure strategy and design. This is due to the fact that the structure and each of its components have been created to coexist peacefully and accommodate one another. Code regulations may need to be reevaluated when photovoltaic systems are added after construction is complete to ensure that the solar power system may be easily included while still being functional and complying to the goals of the codes. While the specifics of those codes may vary from region to region, anyone installing solar panels may generally encounter a few issues, such as ambiguous or constrictive language usage in the codes and a lack of knowledge due to the disparity between new PV models, practices, and technologies and the unupdated version of the codes. varied states, municipalities, and jurisdictions have varied laws and regulations, and the inconsistent, inaccurate, and ineffective nature of code enforcement can also lead to a number of problems.

Remember that some states adopt the codes and regulations on a state-wide basis, while others permit each area to follow its local adaptation of the electrical, building, and fire codes, laws, and regulations when installing solar panels, especially if you are actually doing it yourself. It could be confusing if, for example, your state demands that you have a rooftop area of at least 200 square feet before you can use a solar power system, but if you reside in a certain area, you might be able to install panels with only 150 square feet of space.

The beautiful thing about state-level codes is that they provide you a sense of comprehension and uniformity. Retailers and businesses benefit from this uniformity as well, which results in reduced costs compared to jurisdictions that permit local adaption of the legislation. You virtually ever find fair prices with this variety and inconsistent pricing. The problem, though, would be that new knowledge and practices may not be updated as quickly in the state-wide codes. The terms "Dillon's Rule" and "Home Rule" are very formal names for these alternating adaptations. The states who choose the former option are the ones that have all the power unless particular areas are expressly delegated to the local governments. In the latter states, local jurisdictions are given complete authority, with the exception of those that are specifically mentioned as falling under state authority.

Authorities Having Jurisdiction, or AHJs, are frequently distinct cases. In terms of the steps involved in installing solar panels, these states often have various and occasionally overlapping powers and responsibilities. To be more explicit, Authorities Having Jurisdiction, which may include the state's health departments, county building, and fire marshals, are in charge of enforcing the building and fire rules. State laws must be followed by local governments in states that adhere to Dillon's Rule. Home Rule states are permitted to enact laws that apply both statewide and locally.

A critical step in the installation process is code enforcement. Major catastrophes may occur if regulations are not adhered to and instructions are not rigorously carried out. You must keep in mind that even the most comprehensive and unambiguously stated codes might not be able to serve their intended purpose if any of the parties involved—such as building inspectors, permit staff, and fire marshals—doesn't have the necessary training to be consistent, helpful, and accurate in applying the codes and adhering to their standards. The implication of this is that if you intend to build your own solar panels, you must be extremely mindful of and knowledgeable about what you're doing. If you don't, not only do you run the chance of destroying your new solar power system, but you also run the risk of endangering your neighborhood, your community, and yourself. Local authorities are always in charge of enforcing any kind of legislation and adoption of codes. It can be very difficult to understand the regulations and legalities governing solar power systems and the installation of any kind of solar panel.

Model Codes and Their Location

Local authorities' ability to provide local governments with a fair amount of flexibility and access to local expertise is very beneficial, but the variations among different jurisdictions can lead to needless delays, excessive permitting costs, and increased compliance costs for solar installers. Furthermore, because of these discrepancies, expert inspections may also be inaccurate. Any solar installer that works in multiple jurisdictions will tell you that this is a common problem. It is also an issue for state administrators who are in charge of managing implementation plans for the whole state.

Policymakers and state-level administrators charged with overseeing solar programs must ensure that they provide direction and resources that support local governments in promoting a level of consistency when it comes to enforcing and adopting codes related to photovoltaic installation in order to help provide a workable solution for this problem.

To promote the lowest standards and correspondence among Authorities Having Jurisdiction, numerous model codes have been created. The most important of them is the International Code Council, a group that promotes safety and fire prevention. Every three years, it is eager to offer updates for all of its models. The International Code Council created the International Building Code, International Fire Code, and International Residential Code as illustrative model codes. With municipal and state-level adjustments, these codes have been widely adopted throughout the United States. According to a National Renewable Energy Laboratory paper titled "Clean Energy in City Codes: A Baseline Analysis of Municipal Codification across the United States," 45% of the communities in a sample of 1,266 have references to solar energy in their municipal rules. In addition, this analysis discovered that 30% of the municipal codes' references to solar energy in the four states that were chosen—Florida, Minnesota, California, and Maryland—relate to criteria for design and construction. Most of the codes declared that:

Solar equipment needs to be hidden from view

The Uniform Solar Energy Code and other safety and fire regulations must be followed.

It must adhere to a wide range of design specifications, including the neighborhood's general landscape, aesthetic, and building orientation.

Setback requests must be addressed

Be aware that the analysis indicated that many references omitted to discuss the required building height for solar installations. For your information, this study was released in December 2016. Look through the preceding scripts to make sure you don't encounter any issues. Inquire with your local government about their position on off-grid solar power systems. This will make sure you're headed in the proper direction.

Legalities of Off-Grid Solar Panels

You might be startled to learn that homeowners who use solar electricity and live on the grid are more likely to face legal issues than those who do not. Being off the grid enables you to produce your own electricity while preventing unneeded problems with the government. Those who use the grid could be subject to national standards as well as minimum consumption fees. Despite the fact that the National Electric Code is generally relevant to electrical installations, as someone who will be using off-the-grid solar panels, you need to be aware that DC systems with low voltage have a lot less rigorous limitations. This also applies to mobile homes and tiny houses. This makes it considerably simpler to create modular off-grid power solutions for anyone looking for a short-term solar fix. When it comes to intricate regulations, these are essentially nonexistent. The authorities won't hold it against you if you decide to run USB cables rather than utilizing extension cords, even if it is not actually advised. The services and costs that solar power and installation businesses provide for on-the-grid solar panels are also significantly impacted. They might overcharge individuals who are connected to the grid in an effort to actually push back against the strict regulations; as a result, they would be forced to pay expensive minimum use fees. When that happens, even though people are their own power sources, they still have to pay when they use a lot of electricity. Anyone who produces the bulk, if not all, of the power they require will find this to be very unfair. Off-the-grid solar power consumers are not now compelled by law to pay anything. If your location has publicly owned utilities, you need still be cautious about any potential required hook-ups.

You must stay as much as possible on the low voltage side of the DC system if you want to operate off-the-grid installations within the bounds of the law. Use just 'temporary' wiring, please. If not, you might be implying a long-term installation. Being off the grid necessitates permanent disconnection from the grid; avoid tying your property to the grid. Off-grid living is a difficult step to take. When those connected to the grid exhaust their supply of self-generated energy, they can draw power from the neighborhood utility. On the other side, you are unable to do the same. This is why you need to make sure that you are self-sufficient and that you have looked into everything that could prevent you from using solar energy. There are various laws that you need to take into account before the installation procedure if you want to get the most of your solar power system.

National Electrical Code

The National Electric Code is among the most crucial codes that you should review. Public access to the NEC's rules governing off-grid solar power systems is available online. You can discover detailed and noteworthy requirements for your off-the-grid photovoltaic system in this code. All 50 states consistently enforce the codes found in this resource, regardless of the differences in state laws. If hiring a professional solar installer is something you're considering, make sure they adhere to the NEC. If you are doing it yourself, you might need to carefully read the code or get a professional to assess the installation procedure to make sure you are following the law.

Local Nuisance Regulations

Each city has the authority to issue orders that set restrictions on what landowners can do with or on their properties. Although there may not be a specific city ordinance that forbids the installation of solar panels, there may be restriction concerns that are covered by a local beautification order. As frustrating as it may sound, many localities have vaguely worded ordinances that forbid the installation of anything that appears unusual or unattractive on the roof of your home. The odds are that these regulations could be used to prevent you from producing your own power if your city has certain restrictions, even if solar panels don't generally appear unattractive. However, beauty is a matter of opinion.

Covenants and Homeowner Associations

There are an increasing number of rural land parcels being sold with restrictions or covenants. Bylaws and covenants have been carefully crafted to guarantee that the land you buy is safeguarded from potential deterioration. Several activities, such as owning cattle or even junk equipment, might cause a decline in the value of the property. It's important to notice that the wording of the agreements might be somewhat unclear because there are so many possible things they could say. Simply said, the rules are uncertain. If there are any restrictions, such as ones on the size of your specific solar power system or the number of solar panels you can install, you can find out by checking them out.

RVs and off-grid living

Probably no regulations exist that forbid RV owners from adding solar panels to their vehicles. Remember, nevertheless, that laws and rules pertaining to driving safety still very much apply. You may be unsure of how safe it is for you to drive about town with solar panels on your RV's roof. You must make sure that the panels are very securely fastened to protect both you and others safe. When you accelerate a little, they shouldn't move around or come off, creating a safety concern on the road. Another issue to think about is the potential for your solar panels to reflect sunlight into other drivers' eyes, potentially injuring or momentarily distracting them.

Does having power make sense?

Many individuals wonder whether they need to have power in their houses all the time when they consider installing solar panels and getting completely off the grid. Even while it might seem like a private matter, several states actually do require you to have electricity in your house. Simply put, the answer to this question depends on the region or state in which you reside. There are numerous contexts when having no or little access to electricity is unquestionably acceptable (see the Amish). Particularly for persons who reside in homes that they own, some states don't expressly address this issue in their regulations. However, you must keep in mind that practically every city has strict laws governing the protection of tenants' rights. This means that essential utilities, usually including power, should always be provided by landowners. There are typically no explicit requirements if you genuinely own and reside in your home, presuming that everyone would wish to have access to power where they live. Instead, you will find laws that specify where switches and outlets should be placed, safe ways to run wire, and more. As long as you don't whine about the lack of electricity if you run out, you should be good to go.

Solar panels off the grid and permissions

Depending on where you reside, you may or may not actually require a permission to install an off-the-grid solar power system. As you may remember, every state, county, and city has a different strategy for practically everything we do, including the installation of solar panels. Regarding off-the-grid solar panel installations and whether you would need a permission to put one on your roof, each of them has different requirements. But in most places, especially if you're using a DIY photovoltaic kit, you won't actually need to get a permit. In order to learn more about this matter, it is still advised that you get in touch with your local authority. Remember that rural areas are typically less regulated than urban ones, wherever in the world. This indicates that you are more likely to need a permit to start constructing if you live in an urban area of the city. If you discover that you require one, you may either apply for one using solar permit providers or handle it yourself. To assist you in estimating the cost of obtaining a permit, you can discover free solar quote tools online.

Facts to Consider

Typically, no permit is needed for anyone installing solar panels that face away from the street. You will need to learn more about your area's codes and regulations if you are placing it on a roof that faces the street because you can be in violation of a beautifying ordinance. If your building is heritage listed, you should definitely consult the local council before making any changes to the exterior of your structure. Even if your house is very recent and you live in a heritage-listed district, you still need to inquire ahead of time. Generally speaking, you are not required to let your neighbors know about your plans to install solar power. The installation process could be noisy, so it would be wonderful if you could let them know what you intend to do.

In general, there are several laws, codes, and regulations governing the installation of solar panels. The restrictions and limitations do, however, lessen as you move toward possibilities for off-the-grid solar power systems. Despite this, you still need to be sure that you are truly abiding by the laws in your area because each state, county, and city has its own set of norms and regulations.

Solar Power FAQs

What actually distinguishes solar thermal energy from photovoltaic energy?

Collectors use solar thermal energy to transform sun radiation into useful heat. Heat exchangers are typically used to prepare hot water and heat it. Direct solar energy conversion into electricity is referred to as photovoltaic.

What Stands for Photovoltaics?

The direct conversion of solar energy into electricity using solar modules composed of individual solar cells is known as photovoltaics. The word is derived from the Greek words photon, which means light, and volta, an Italian inventor who gave his name to the volt unit of electrical voltage.

How do solar cells work?

A solar cell is typically a small, silicon disc with dimensions of 15 x 15 cm. Contact strips, which are attached to the front and back, transport the current produced by the solar cell's current away. Sand derived from quartz is used to make silicon. Less frequently used are other crystalline and thin-film solar cells. Less than 10% of the market is taken by both of them.

How long does it take to install a PV system?

A solar system's installation time is influenced by its capacity and structural circumstances. The installation process takes one to three working days for systems up to ten kWp.

Can a Flat Roof Also Have a Photovoltaic System Installed?

PV installations are straightforward on flat rooftops. In this instance, the modules are mounted using a mounting mechanism at an angle to position them with a good slope and optimal south-facing. In addition, the modules' slanted installation ensures proper ventilation and cleaning results. Here, it's crucial that the modules are spaced apart sufficiently to prevent shading and a subsequent drop in yield. The flat roof area needed to install the same output is almost twice as large as with a pitched roof alternative.

What Distinguishes an Island System from a Grid-Connected Solar Power System?

In a solar power system that is connected to the grid, some or all of the power produced by the sun is fed into the national power grid. In a stand-alone system, the electricity produced is consumed entirely by the business and is not fed into the general power grid. There is no electrical grid connection for island systems.

What Styles of Mounting for Photovoltaic Systems on Roofs Are Available?

On-roof and in-roof installation are differentiated in a fundamental manner. The PV modules are mounted to mounting points above the preexisting roof covering in an on-roof installation. For flat roofs, a mounting mechanism is used to set up the modules at an angle so that they are aligned at an angle.

The photovoltaic modules are embedded into the roof during in-roof installation. The roof tiles are replaced by solar roof tiles, which are plugged together on the underside to form larger units. The unique solar panels and more difficult installation make in-roof mounting more expensive despite its better aesthetics.

What Does Wp Stand for in a Power Specification?

Watt peak is abbreviated Wp. In photovoltaics, this is a unit that is well-established. It represents the maximum power that a solar module can produce at noon under ideal lighting circumstances (among other factors).

How Much Electricity Is Produced by a PV System?

The position, inclination, orientation, and solar module configuration all affect how much energy is produced by the sun. A solar system typically produces 1,000 kilowatt hours of power per kWp, with the southern part frequently producing somewhat more electricity than the northern portion.

What About Overcast Sky?

Photovoltaic modules use both diffuse radiation that travels through clouds as well as direct sunlight, which is only present when the sky is clear. Whether or not the sun is directly visible, the modules create more power the brighter it is. How Safe and Durable is the System? About half of the radiation is dispersed, incidentally.

The reliability of photovoltaics has been established over many years and in a wide range of applications. Since the 1960s, photovoltaics have been utilized successfully in space flight. Since the 1990s, tens of thousands of small and big systems have been constructed. An annual decline in module output of between 0.3 and 0.5 percent is now anticipated with a service life of around 30 years. The cover glass, metal frame, and partially the solar cells can then be recycled, for instance to make new solar modules.

Isn't There More Energy Used in Production Than the Solar System Can Produce?

The solar system must create energy for around two years in order to generate the energy that will be put back into it. It will consequently create more than 10 times as much clean power as was required to produce it during the course of its service life, which must be at least 25 years.

What Else Is Included in a PV System?

The following parts make up a photovoltaic system; they are individually matched to one another and so tailored to the fundamental structural requirements of the home:

- solar cells that use sunlight to produce direct current.
- Inverters that change direct electricity into alternating current that is compatible with the grid.
- a feed-in meter that calculates the amount of electricity pumped into the grid.
- The solar modules' mounting structure, wiring, and wiring.
- Power storage that temporarily stores the electricity produced by the sun.

Where ought the inverter to be mounted?

The inverter should ideally be mounted in a protected area that stays cool even in the summer. There are inverters that can also be installed outside if there isn't an appropriate location within the home.

Does a PV system require routine cleaning?

Not frequently, as rain and snow usually take care of cleaning due to the smooth surface of the module.

Does the Photovoltaic System Need a Permit?

The grid connection needs to be clarified before the installation is put into service because the grid operator may have additional requirements, particularly for bigger systems over 30 kW. Additionally, taking monument protection into account and, in rare cases, getting a building permit may be necessary. Even if this shouldn't typically be required, it doesn't harm to inquire with the local government or office just to be safe.

The Federal Network Agency must be informed of the system's location and functioning following completion. If the report is ignored, penalties may be implemented after the fact.

How is Electricity Introduced into the Grid?

The solar system is connected to the typical 230-volt network after the direct current generated in the modules is converted into the typical mains alternating current via an inverter. The electricity produced by the photovoltaic system is always used, either in your home or by homes nearby that are connected to the local power grid.

What Size Solar System Is Appropriate for a Family?

If no power is used for heating or hot water, a family of four might anticipate to use a total of 3,500 to 4,500 kWh of electricity annually. For this amount of electricity to be produced, a module size of 30–40 m2 is needed, depending on the type of cell. This generally translates to a 5 kWp system output. Naturally, depending on the needs of the customer, a higher output may also be used if the roof surface permits it.

Photovoltaics: Are They Too Expensive?

The underlying system is regarded as mature, and prices have recently decreased dramatically as a result of pressure from worldwide competition. Further price drops for solar modules, however, are so negligible that they are nearly entirely offset by labor costs for installers, which have risen over time.

Exist Any Funding Options?

The Reconstruction Loan Corporations offer subsidized loans in every state. Additionally, some localities have regional funding. Asking at the Citizens' Office or regional office is worthwhile.

Is Purchasing a Photovoltaic System Still Worth It?

PV systems are worth the expenditure, especially for self-consumption. Self-generated electricity is typically less expensive than electricity purchased from the public grid.

Additionally, you get a feed-in tariff for 20 years that is state-guaranteed for any electricity that is not used but instead is put into the public grid.

Is a Surcharge Required for Personal Use?

You'll be charged a surcharge for the portion of the electricity you consume personally. For systems with more than 10 kWp, this is applicable. The majority of systems installed on single-family homes are free from the surcharge because they produce less than 10 kWp.

Is it Possible to Affect the Photovoltaic System's Profitability?

The system's profitability can be increased by using as much of the self-produced electricity as feasible, as this profitability grows as self-consumption increases. Consequently, use more electricity during the day than at night. A battery for daily temporary storage can be a way to boost self-consumption if this is not possible owing to work, for instance.

Does it matter if the real roof does not face south if a PV system is installed?

Given that most of the electricity produced is used when a person is at home, an east-west orientation may make more sense. Because the sun rises in the east and sets in the west, an east-west orientation is particularly appealing to people who work during the day.

Does it Make Sense to Purchase a Photovoltaic System Today and to Add Battery Storage Later?

Battery storage can be added to a photovoltaic system at any time. The installation of a storage and PV system at the same time, however, is less expensive due to reduced system costs (combination devices are less expensive), as well as lower installation costs if the craftsman does not need to make two trips.

Are the Feed-In Earnings Taxable?

The revenue generated by a solar system must be considered for tax purposes and submitted to the appropriate tax authority. Consult your tax expert to get the specifics.

The electricity must be purchased by the grid operator, right?

Yes, the network operator is required by law to buy the electricity, and priority connection and compensation are also subject to legal restrictions.

Conclusion

Renewable energy sources are in high demand as the globe looks for alternative energy sources to help heal the planet's atmosphere. To counteract the escalating temperatures, every country is using a distinct set of tactics. Finding alternatives to fossil fuels is more crucial than ever, especially as the energy problem becomes more and more severe.

In addition to being renewable, solar energy is also widely accessible. Until the sun dies, which is expected to take another 14 billion years, humanity can rely on solar energy every day. The ecology benefits from the use of solar energy. Most solar panels live longer than that, often between 15 and 25 years. It is only a matter of time before the world develops solar technology for nearly every purpose, especially with the way technology is developing and solar power materials becoming more affordable all the time.

Photons are used in solar panels to cause electron motion, which produces energy. The energy is converted from direct current into alternating current, which is the energy type most frequently utilized in large appliances and equipment. This conversion requires an inverter. Always utilize correct racking when adding solar panels to any flat surface. The solar panel could easily be damaged in bad weather if improperly installed. The charge controller controls how much power is sent from the solar panels to the batteries. The energy delivered into the batteries could overload without a charge controller, rendering them worthless.

If you reside in an area with year-round sunlight, you might be able to power your entire house or mobile equipment with just a small number of panels, reducing your energy costs. As solar energy technology advances, it now also powers lighting and heating systems as well as recharging wearable electronics.

But there is a downside to every upside. Installing solar panels is frequently pricey at first. Even though they require relatively little care, after the initial installation you could feel a pinch in your wallet. Additionally, the weather greatly affects solar panels. On gloomy days, photons may still pass through clouds, but not nearly as effectively. Therefore, solar panels might not be the best option for you if you reside in a region that experiences a lot of rain. Batteries are not the least expensive option, and if you need to change them frequently, they can cost quite a bit. However, this is less frequent with smaller installations. Wildfires have also been caused by solar panels by physically setting wildlife on fire. Solar panels are also challenging to disassemble. They contain the hazardous material cadmium, which when it rains on damaged panels can leak into the environment. Solar panels still harm the environment if not disposed of correctly until a safer method is developed.

Use a stand to hold the solar panels and batteries when creating a mobile solar power station. These are some of the simplest configurations to construct, and they can power devices everywhere. It's also common and getting simpler all the time to mount solar panels on automobiles and recreational vehicles. Solar panels frequently need more care and upkeep because many people spend hundreds of hours traveling each year. Last but not least, putting solar panels on a boat demands careful consideration. Although installing the panels on a yacht is just as simple as installing them anywhere else, they demand additional care. Except for the solar panels, all equipment must be stored below deck to protect it from moisture and salt, which could cause it to corrode.

Use your imagination when constructing your own system. The ideas presented in this book offer recommendations and counsel, but how they are constructed is entirely up to you. When working with any materials, keep in mind to compare prices. If you don't look around, you might not realize the hidden gems the solar panel market has to offer.

ACCESS YOUR FREE VIDEO COURSE

Complete video course on how to understand, install, run and maintain your perfect Solar System

19 videos to help you build your DIY Off-Grid Solar System in the easiest, safest and fastest way. Includes step by step explanations and best practices.

Scan the QR code now and start your learning.